Atomicity through Fractal Measure Theory

Alina Gavriluţ • Ioan Mercheş • Maricel Agop

Atomicity through Fractal Measure Theory

Mathematical and Physical Fundamentals with Applications

Springer

Alina Gavriluţ
Faculty of Mathematics
Alexandru Ioan Cuza University
Iaşi, Romania

Ioan Mercheş
Faculty of Physics
Alexandru Ioan Cuza University
Iaşi, Romania

Maricel Agop
Physics Department
Gheorghe Asachi Technical
University of Lasi
Iaşi, Romania

ISBN 978-3-030-29595-0 ISBN 978-3-030-29593-6 (eBook)
https://doi.org/10.1007/978-3-030-29593-6

Mathematics Subject Classification: 28B20, 28E05, 28A80, 32A12, 81Q35, 81P15, 28C20

This Springer imprint is published by the registered company Springer Nature Switzerland AG.
The registered company address is: Gewerbestrasse 11, 6330 Cham, Switzerland

"Gravitation is not responsible for people falling in love"

(Albert Einstein)

Preface

Any material structure is theoretically conceived in particle terms. These particles are in no way "elementary" but show themselves a structure in agreement with the material structure to which they temporarily belong. In short, neither that material structure nor the structure of the constitutive particles is fixed: they are essentially ephemeral, being on a continuous transformation. The classical case of the molecules, atoms, and nuclei is mostly known: all of these structural particles finally showed to be different as compared to the initial belief. Closer to nowadays, the subject elementary particles presents a large variety of terms that make obsolete the initial meaning of "elementary" denomination.

But the classical force theory (to some extent) and the quantum force theory (to a greater extent) are not able to really face this emergent imago of the world—which in our opinion is quite natural—to conceive the particle as a material point. This image can be fruitfully used only for big distances between particles, but it fails if the interacting particles are closed to each other. In this last case, the classical material point has to be replaced by an extended particle, and this book sets out the basis for this perspective.

We make the observation that by an extended particle we mean a finite, spatially extended material volume and *a fortiori* a limit of matter separation with respect to the empty space. It seems natural the fact that, when incorporated in a material structure, an extended particle has to suffer transformations in order to face that structure. This means that, structurally, its internal evolution has to temporarily belong to the evolution of the corresponding material structure. This observation puts into evidence the idea of "envelope" (interface) or, differently speaking, the separation limit of the particle. Therefore, our purpose is to describe the separation limit from a mathematical perspective, emphasizing the possibilities of introducing physics into this mathematical theory. In this respect, we show that approaching his investigation by means of the quantum measure theory can satisfactorily accomplish the physical necessities. More than that, taking into account that in such a context typical quantum mechanical paths (as shown by Feynman) are fractal curves with fractal dimension 2, and since probability is an important notion in quantum mechanics, a holographic perspective by means of a fractal measure theory can be approached. Measure theory's technique could be used to study quantum phenomena. Unfortunately, one of the foundational axioms of measure theory does not remain valid in its intuitive application to quantum mechanics. Modification of the traditional measure theory led to quantum measure theory. Here an extended notion of a measure has been introduced.

Since additivity—a foundational axiom of measure theory—fails in quantum mechanics, weaker variants of additivity (null-additivity, null-null-additivity, null-equality, etc.) have been proposed to be what is called today the quantum measure theory.

So, in mathematical (and not only) recent literature, due to many theoretical and practical necessities, the classical problems from measure theory were first transferred to submeasures. Since subadditivity is still a strong condition, the study of other set of functions having important asymptotic structural properties (much less restrictive than subadditivity), such as null-additivity, null-null-additivity, etc. (which we have already mentioned) started to be developed. Especially, continuity properties (such as continuity from above/below order continuity, regularity) have been studied intensively due to their various applications in economics, game theory, artificial intelligence, and many other important fields.

In fact, topological methods have been used for many years in the study of the chaotic nature of dynamical systems. Most of these systems are collective phenomena emerging out of many individual components. Therefore, a need for a topological study of these collective dynamics arose. Precisely, it became more and more clear that problems treated in the single-valued setting need to be extended in the set-valued measure framework. This set-valued general setting requires the use of hypertopologies (Hausdorff, Vietoris, Wijsman, Fell, Attouch-Wets, and many others) and has become of great interest in matters connected to optimization, convex analysis, the economy, etc.

On the other hand, because of multiple applications (game theory, mathematical economics), a study concerning (non)additive set (multi)functions has been recently developed. The notion of an atom and its variants (pseudoatom, nonatomicity, non-pseudoatomicity, finite purely atomicity) plays a key role in measure theory and its applications and extensions.

The present book is organized as follows. In Chapter 1, a short overview of several well-known hypertopologies is made. In Chapter 2, we present a unifying mathematical point of view on hypertopologies and regularity in the context of fuzzy set multifunctions. Certain relationships, interpretations, and similitudes are thoroughly highlighted. Also, in these two chapters, certain main notions that will be used throughout the book are given. In Chapters 3, 4, and 5, various problems concerning atoms/pseudoatoms are discussed for fuzzy set multifunctions taking values in the family of all nonvoid closed subsets of a Banach space in Hausdorff topology. In Chapter 6, results referring to Gould-type integrability on atoms are presented for monotone set multifunctions taking values in the family of all closed nonempty subsets of a Banach space, family that is endowed with the Hausdorff topology. In Chapter 7, some continuity properties, such as increasing/decreasing convergence, exhaustivity, order continuity, regularity, are introduced and studied in

Vietoris topology for fuzzy set multifunctions taking values in the family of subsets of a Hausdorff linear topological space. In Chapter 8, we continue the study began in Chapter 7 concerning continuity properties for set multifunctions taking values in all nonvoid subsets of a linear topological space. Based on these results, Egoroff and Lusin-type theorems are obtained and Lusin-type theorems are obtained for a set of multifunctions in Vietoris topology. In Chapter 9, atomicity is discussed via regularity for set multifunctions taking values in the family of all nonvoid subsets of a topological space. In Chapter 10, atomicity is presented via quantum measure theory, and some of its physical applications are highlighted. Precisely, the mathematical concept of (minimal) atomicity is extended from a physical perspective based on the nondifferentiability of motion curves. In Chapter 11, a multifractal theory of motion is built up and, moreover, a possible multifractal theory of measure is proposed.

For the reader's convenience, each chapter is conceived separately and can be read independently, without reference to the other chapters.

All results contained in this book are based on recent researches (individual or joint) of the authors, which results are either already published or submitted for publication. The study started in these papers will surely be extended in some further research.

The book is addressed to graduate and postgraduate students, teachers, and all researchers in physics and mathematics but not only. Since the notions and results offered by this book are closely related to various notions of the theory of probability, we wish that this book would be useful to a wide category of readers, using multivalued analysis techniques in areas such as control theory and optimization, economic mathematics, game theory, etc. Also, it is assumed that the reader is familiar with the basic concepts and results from measure theory and/or theoretical physics.

As a personal note, we would like to thank our families for their support and understanding.

Iaşi, Romania
2019

Alina Gavriluţ
Ioan Mercheş
Maricel Agop

Contents

Chapter 1
Several hypertopologies: A short overview

1.1 Introduction

In the recent decades, due to its various applications, the study of hypertopologies has become of a great interest. The idea of modelling the behaviour of phenomena at multiple scales has become an useful tool in pure and applied mathematics. Fractal-based techniques lie at the heart of this area, since fractals are multiscale objects, which often describe such phenomena better than traditional mathematical models do. In Kunze et al. [19] and Wicks [25], certain hyperspace theories concerning the Hausdorff metric and the Vietoris topology are developed as a foundation for self-similarity and fractality.

For many years, topological methods were used in many fields to study the chaotic nature in dynamical systems (Sharma and Nagar [23] and many others), which seem to be collective phenomenon emerging out of many segregated components. Most of these systems are collective (set-valued) dynamics of many units of individual systems. It therefore arised the need of a topological treatment of such collective dynamics. Recent studies of dynamical systems, in engineering and physical sciences, have revealed that the underlying dynamics is set-valued (collective), and not of a normal, individual kind, as it was usually studied before. An interesting approach of topology in psychology was given by Brown in [9]. We also note different considerations concerning generalized fractals in hyperspaces endowed with the Hausdorff hypertopology, or, more generally, with the Vietoris hypertopology (Andres and Fiser [1], Andres and Rypka [2], Banakh and Novosad [5], Kunze et al. [19]). Interesting results were obtained by Lorenzo and Maio [11] in melodic similarity, Kunze et al. [19], Wicks [25] in self-similarity and fractality.

Important theoretical results concerning hypertopologies can be found in Beer [6, 7], Beer et al. [8], Apreutesei [3, 4], Hu and Papageorgiou [18], Frolík [13], Precupanu et al. [21], Precupanu et al. [22], Hazewinkel [17], Solecki [24], Wijsman [26], etc.

A. Gavriluţ et al., *Atomicity through Fractal Measure Theory*,
https://doi.org/10.1007/978-3-030-29593-6_1

Wijsman proposed in [26] the weak topology on the collection of nonempty closed subsets of a metric space generated by the distance functionals viewed as functions of set argument. Researchers in hypertopologies found this topology quite useful and subsequently a vast body of literature developed on this topic since Wijsman topology describes better the pointwise properties. For instance, in some examples of fractals, like neural networks and circulatory system, the uniform property of the Hausdorff topology is inappropriate.

In this chapter, for the convenience of the reader, we list the definitions and main properties of three important hypertopologies, Hausdorff, Vietoris and Wijsman. In the following chapter, we shall present a more detailed overview concerning several main hypertopologies and also related considerations concerning regularity with respect to these hypertopologies.

1.2 Vietoris and Hausdorff topologies

Allover this book, by $\mathbb{N}$ we denote the set of all naturals and by $i = \overline{1, n}$ we usually mean $i \in \{1, 2, \ldots, n\}$. By $\mathbb{N}^*$ we mean $\mathbb{N}\backslash\{0\}$. Also, $\mathbb{R}_+ = [0, \infty)$ and $\overline{\mathbb{R}}_+ = [0, \infty]$.

Let (X, τ) be a Hausdorff, linear topological space with the origin 0. $\mathcal{P}_0(X)$ denotes the family of all nonvoid subsets of X.

We recall the following basic notions and notations (Hu and Papageorgiou [18, Ch. 1], Precupanu et al. [22, Ch. 1], Precupanu et al. [21, Ch. 8]):

$M^- = \{C \in \mathcal{P}_0(X);\, M \cap C \neq \emptyset\}$, $M^+ = \{C \in \mathcal{P}_0(X);\, C \subseteq M\}$,
$S_{UV} = \{D^+;\, D \in \tau\}$ (UV-for upper Vietoris topology) and
$S_{LV} = \{D^-;\, D \in \tau\}$ (LV-for lower Vietoris topology).

Vietoris topology, denoted by $\widehat{\tau}_V$, is the topology on $\mathcal{P}_0(X)$ which has as a subbase the class $S_{UV} \cup S_{LV}$.

$\widehat{\tau}_V^+$ - *the upper Vietoris topology* (respectively, $\widehat{\tau}_V^-$ - *the lower Vietoris topology*) denotes the topology which has as a subbase the class S_{UV} (respectively, S_{LV}). Obviously, $\widehat{\tau}_V = \widehat{\tau}_V^+ \cup \widehat{\tau}_V^-$.

For $U, V \in \tau$, define $\mathcal{B}_{U,V_1,V_2,\ldots,V_k} = U^+ \cap V_1^- \cap V_2^- \cap \ldots \cap V_k^-$. Then the family $\mathcal{B}_{U,V_1,V_2,\ldots,V_k}$ of such subsets, where $U, V_1, V_2, \ldots, V_k \in \tau$, is a base for the topology $\widehat{\tau}_V$.

Also, the family of subsets $\mathcal{B}_U = U^+$ (respectively, $\mathcal{B}_V = V^-$) is a base for $\widehat{\tau}_V^+$ (respectively, $\widehat{\tau}_V^-$).

If (X, d) is a metric space, $\mathcal{P}_0(X)$ is the family of all nonvoid sets, $\mathcal{P}_b(X)$ denotes the family of all nonvoid bounded subsets of X, $\mathcal{P}_f(X)$ denotes the family of all nonvoid closed subsets of X, $\mathcal{P}_{bf}(X)$ is the family of nonvoid closed bounded subsets of X. $\mathcal{P}_k(X)$ is the family of all nonvoid compact subsets of X.

Let be τ_d the topology induced by d and h, the Hausdorff-Pompeiu pseudometric on $\mathcal{P}_0(X)$.

h becomes a veritable metric on $\mathcal{P}_{bf}(X)$ (Hu and Papageorgiou [18, Ch. 1], Precupanu et al. [22, Ch. 1], Precupanu et al. [21, Ch. 8]).

If X is a Banach space, then $(\mathcal{P}_f(X), h)$ is a complete pseudometric space. We recall that

$$h(M, N) = \max\{e(M, N), e(N, M)\}, \quad \text{for every } M, N \in \mathcal{P}_0(X),$$

where

$$e(M, N) = \sup_{x \in M} d(x, N),$$

(the *excess of M over N*) and $d(x, N)$ is the distance from x to N with respect to d ($d(x, N) = \inf_{y \in N} d(x, y)$).

Also

$$h(M, N) = \sup_{x \in X}\{|d(x, M) - d(x, N)|\}.$$

If $\varepsilon > 0$, $a \in X$ and $M \subset X$, we denote by

$$S(a, \varepsilon) = \{x \in X; d(x, a) < \varepsilon\}, S_\varepsilon(M) = \{x \in X; \exists a \in M, d(x, a) < \varepsilon\}.$$

Hausdorff topology $\hat{\tau}_H$ on $\mathcal{P}_f(X)$, induced by the Hausdorff pseudometric h can also be interpreted as the supremum of the *upper Hausdorff topology* $\hat{\tau}_H^+$ and the *lower Hausdorff topology* $\hat{\tau}_H^-$ (Hu and Papageorgiou [18, Ch. 1]).

If $M \in \mathcal{P}_f(X)$, then a base of neighbourhoods for $\hat{\tau}_H^+$ is $\{U_+^H(M, \varepsilon)\}_{\varepsilon > 0}$, where $U_+^H(M, \varepsilon) = \{N \in \mathcal{P}_f(X); N \subset S_\varepsilon(M)\}$.

A base of neighbourhoods for $\hat{\tau}_H^-$ is $\{U_-^H(M, \varepsilon)\}_{\varepsilon > 0}$, where $U_-^H(M, \varepsilon) = \{N \in \mathcal{P}_f(X); M \subset S_\varepsilon(N)\}$.

Now, suppose that, moreover, $(X, \|\cdot\|)$ is a real normed space. By $\mathcal{P}_c(X)$ we denote the family of all nonvoid convex subsets of X and by $\mathcal{P}_{bfc}(X)$, the family of nonvoid closed bounded convex subsets of X. $\mathcal{P}_{kc}(X)$ denotes the family of compact, convex subsets of X.

For every $M \in \mathcal{P}_0(X)$, we denote $|M| = h(M, \{0\})$.

For the convenience of the reader, we list several important properties of e, h and $|\cdot|$. Some of them will be used throughout this book:

(i) $h(\overline{M}, \overline{N}) = h(\overline{M}, \overline{N})$ (where $\overline{M}$ is the closure of M with respect to the topology induced by the norm of X), for every $M, N \in \mathcal{P}_0(X)$;

(ii) $e(M, N) = 0$, so $e(N, M) = h(M, N)$, for every $M, N \in \mathcal{P}_f(X)$, with $M \subseteq N$;

(iii) $e(M, N) \leq e(M, P) + e(P, N)$ and $h(M, N) \leq h(M, P) + h(P, N)$, for every $M, N, P \in \mathcal{P}_f(X)$;

$h(M_1 + M_2, N_1 + N_2) \leq h(M_1, N_1) + h(M_2, N_2)$, for every $M_1, M_2, N_1, N_2 \in \mathcal{P}_f(X)$;

(iv) If $M, N \in \mathcal{P}_f(X)$, with $M \subseteq N$, then $|M| \leq |N|$;
(v) $|M + N| \leq |M| + |N|$, for every $M, N \in \mathcal{P}_f(X)$;
(vi) $h(M + P, N + P) \leq h(M, N)$, for every $M, N, P \in \mathcal{P}_f(X)$;
Also, $h(M + P, N + P) = h(M, N)$, for every $M, N, P \in \mathcal{P}_{kc}(X)$;
(vii) $\alpha M \subseteq \alpha N$, for every $M, N \in \mathcal{P}_f(X)$, with $M \subseteq N$ and every $\alpha \in \mathbb{R}$;
(viii) $h([0, a], [0, b]) = |a - b|$, for every $a, b \in \mathbb{R}_+$; $|[0, a]| = a$, for every $a \in \mathbb{R}_+$.

1.3 Wijsman topology

Let (X, d) be a metric space.

We recall (Apreutesei [3, 4], Precupanu et al. [21, 22], Gavriluţ and Apreutesei [15], Gavriluţ and Agop [16], Gavriluţ [14]) the following basic notions and results:

Wijsman topology τ_W on $\mathcal{P}_f(X)$ is the supremum of the upper topology τ_W^+ and the lower topology τ_W^-.

The family

$$\mathcal{F} = \{M \in \mathcal{P}_f(X); d(x, M) < \varepsilon\}_{\substack{x \in X \\ \varepsilon > 0}} \cup \{M \in \mathcal{P}_f(X); d(x, M) > \varepsilon\}_{\substack{x \in X \\ \varepsilon > 0}}$$

is a subbase for τ_W on $\mathcal{P}_f(X)$.

τ_W^- is generated by the family

$$U_-^W(M, x_1, x_2, \ldots, x_n, \varepsilon) = \{N \in \mathcal{P}_f(X); d(x_i, N) < d(x_i, M) + \varepsilon, \text{ for every } i = \overline{1, n}\},$$

τ_W^+ is generated by the family

$$U_+^W(M, x_1, x_2, \ldots, x_n, \varepsilon) = \{N \in \mathcal{P}_f(X); d(x_i, M) < d(x_i, N) + \varepsilon, \text{ for every } i = \overline{1, n}\},$$

where $M \in \mathcal{P}_f(X)$, $\{x_1, x_2, \ldots, x_n\} \subset X$, $\varepsilon > 0$.

In fact, $\tau_W^- = \tau_V^-$—the lower Vietoris topology which is the topology having as a subbase the class $S_{LV} = \{D^-; D \in \tau\}$, where $D^- = \{M \in \mathcal{P}_f(X); M \cap D \neq \emptyset\}$, as recalled in the previous section).

Proposition 1.3.1 *Suppose $\{M_i\}_{i \in I} \subset \mathcal{P}_f(X)$. The following statements are equivalent:*

(i) $M_i \xrightarrow{\tau_W} M \in \mathcal{P}_f(X)$;
(ii) For every $x \in X$, $d(x, M_i) \xrightarrow{p} d(x, M)$ *(pointwise convergence);*
(iii) $M_i \xrightarrow{\tau_W^+} M$ *and* $M_i \xrightarrow{\tau_W^-} M$.

Proposition 1.3.2

(i) $M_i \overset{\tau_W^+}{\to} M$ *if and only if for every* $x \in X$, $\liminf_i d(x, M_i) \geq d(x, M)$ *(i.e., for every* $0 < \varepsilon < \varepsilon'$ *with* $S(x, \varepsilon') \cap M = \emptyset$, *there is* $i_0 \in I$ *so that for every* $i \in I$, *with* $i \geq i_0$, *we have* $S(x, \varepsilon) \cap M_i = \emptyset$*).*

(ii) $M_i \overset{\tau_W^-}{\to} M$ *if and only if for every* $x \in X$, $\limsup_i d(x, M_i) \leq d(x, M)$ *(i.e., for every* $D \in \tau_d$ *with* $D \cap M \neq \emptyset$, *there is* $i_0 \in I$ *so that for every* $i \in I$, *with* $i \geq i_0$ *we have* $D \cap M_i \neq \emptyset$*).*

Now, we recall some important properties of Wijsman topology:

If (X, d) is a complete, separable metric space, then $\mathcal{P}_f(X)$ endowed with the Wijsman topology is a Polish space (Beer [6, 7]). Moreover, the space $(\mathcal{P}_f(X), \tau_W)$ is Polish if and only if (X, d) is Polish.

(X, d) is separable if and only if $\mathcal{P}_f(X)$ is either metrizable, first-countable or second-countable (Levi-Lechicki [20]).

The dependence of the Wijsman topology on the metric d is quite strong. Even if two metrics are uniformly equivalent, they may generate different Wijsman topologies. Necessary and sufficient conditions for two metrics to induce the same Wijsman topology have been found (Costantini et al. [10], Lechicki and Levi [20]).

A natural question arises: what is the supremum of the Wijsman topologies induced by the family of all metrics that are topologically (respectively, uniformly) equivalent to d. It has been shown that the supremum of topologically (respectively, uniformly) equivalent metrics is the Vietoris topology $\widehat{\tau}_V$ (Beer [6, 7], Levi and Lechicki [20], Maio and Naimpally [12]).

1.4 Several comparisons among the three topologies

In this last section, we briefly list some results of comparison among the three hypertopologies.

(i) If the pointwise convergence of Wijsman convergence is replaced by uniform convergence (uniformly in x), then one obtains Hausdorff convergence induced by the Hausdorff pseudometric h.
(ii) Generally, Hausdorff topology $\widehat{\tau}_H$ is finer than Wijsman topology τ_W.
(iii) Hausdorff and Wijsman topologies on $\mathcal{P}_f(X)$ coincide if and only if (X, d) is totally bounded.
(iv) If X is a real normed space, then Hausdorff topology, Vietoris topology and Wijsman topology are all equivalent on the class of monotone sequences of subsets of $\mathcal{P}_k(X)$:

On $\mathcal{P}_k(X)$, $\hat{\tau}_H^- = \hat{\tau}_V^-$, $\hat{\tau}_H^+ = \hat{\tau}_V^+$ and $\hat{\tau}_H = \hat{\tau}_V$ (Hu and Papageorgiou [18, Ch. 1], Precupanu et al. [22, Ch. 1], Precupanu et al. [21, Ch. 8]).

1.5 Kuratowski convergence

Let (X, d) be a metric space. For any $x \in X$ and any sequence of subsets $A_n \subseteq X$, $n \in \mathbb{N}$, we define:

(i) *the Kuratowski limit inferior* (or *lower closed limit*) of A_n as $n \to \infty$:
$\underset{n\to\infty}{Li}(A_n) = \{x \in X; \limsup_{n\to\infty} d(x, A_n) = 0\} = \{x \in X;$ for all open neighbourhoods U of x, $U \cap A_n \neq \emptyset$ for large enough $n\}$;

(ii) *the Kuratowski limit superior* (or *upper closed limit*) of A_n as $n \to \infty$:
$Ls(A_n) = \{x \in X; \liminf_{n\to\infty} d(x, A_n) = 0\} = \{x \in X;$ for all open neighbourhoods U of x, $U \cap A_n \neq \emptyset$ for infinite many $n\}$.

If the Kuratowski limits inferior and superior agree (i.e., they are the same subset of X), then their common value is called the *Kuratowski limit* of the set A_n, as $n \to \infty$, and it is denoted by $Lt_{n\to\infty} A_n$.

Generally, $Li_{n\to\infty} A_n \subseteq \underset{n\to\infty}{Ls} A_n$ and $\underset{n\to\infty}{Li} A_n$, $\underset{n\to\infty}{Ls} A_n$ are always closed sets in the metric topology on (X, d).

Kuratowski convergence is weaker than convergence in Hausdorff metric.

For compact metric spaces X, Kuratowski convergence coincides with both convergence in Hausdorff metric and Vietoris topology.

Example Let A_n be the zero set of $\sin(nx)$:

$$A_n = \{x \in \mathbb{R}; \sin(nx) = 0\}, \quad n \in \mathbb{N}^*.$$

Then A_n converges in the Kuratowski sense to the whole real line $\mathbb{R}$.

References

1. Andres, J., Fiser, J.: Metric and topological multivalued fractals. Int. J. Bifur. Chaos Appl. Sci. Eng. **14**(4), 1277–1289 (2004)
2. Andres, J., Rypka, M.: Multivalued fractals and hyperfractals. Int. J. Bifur. Chaos Appl. Sci. Eng. **22**(1), 1250009 (2012)
3. Apreutesei, G.: Set convergence and the class of compact subsets. An. Şt. Univ. Iaşi **XLVII**, 263–276 (2001)
4. Apreutesei, G.: Families of subsets and the coincidence of hypertopologies. An. Şt. Univ. Iaşi **XLIX**, 1–18 (2003)
5. Banakh, T., Novosad, N.: Micro and macro fractals generated by multi-valued dynamical systems, arXiv: 1304.7529v1 [math.GN] (2013)
6. Beer, G.: Topologies on Closed and Closed Convex Sets. Mathematics and its Applications. Kluwer Academic Publishers Group, Dordrecht (1993)
7. Beer, G.: Wijsman convergence: a survey. Set-Valued Anal. **2**(1–2), 77–94 (1994)
8. Beer, G., Lechicki, A., Levi, S., Naimpally, S.: Distance functionals and suprema of hyperspace topologies. Ann. Mat. Pura Appl. (4) **162**, 367–381 (1992)
9. Brown, S.: Memory and mathesis: for a topological approach to psychology. Theory Cult. Soc. **29**(4–5), 137–164 (2012)

10. Costantini, C., Levi, S., Zieminska, J.: Metrics that generate the same hyperspace convergence. Set-Valued Anal. **1**, 141–157 (1993)
11. Di Lorenzo, P., Di Maio, G.: The Hausdorff metric in the melody space: a new approach to melodic similarity. In: The 9th International Conference on Music Perception and Cognition, Alma Mater Studiorum University of Bologna, August 22–26 (2006)
12. Di Maio, G., Naimpally, S.: Comparison of hypertopologies. Rend. Ist. Mat. Univ. Trieste **22**, 140–161 (1990)
13. Frolík, Z.: Concerning topological convergence of sets. Czechoskovak Math. J. **10**, 168–180 (1960)
14. Gavriluţ, A.: Regular Set Multifunctions. PIM Publishing House, Iaşi (2012)
15. Gavriluţ, A., Apreutesei, G.: Regularity aspects of non-additive set multifunctions. Fuzzy Sets Syst. **304**, 94–109 (2016)
16. Gavriluţ, A., Agop, M.: A mathematical-physical approach on regularity in hit-and-miss hypertopologies for fuzzy set multifunctions. Math. Sci. **9**, 181–188 (2015)
17. Hazewinkel, M.: Encyclopaedia of Mathematics, Supplement III, vol. 13. Kluwer Academic Publishers, Dordrecht (2001)
18. Hu, S., Papageorgiou, N.S.: Handbook of Multivalued Analysis, vol. I. Kluwer Academic Publishers, Dordrecht (1997)
19. Kunze, H., La Torre, D., Mendivil, F., Vrscay, E.R.: Fractal Based Methods in Analysis. Springer, New York (2012)
20. Lechicki, A., Levi, S.: Wijsman convergence in the hyperspace of a metric space. Boll. Unione Mat. Ital. (7) B.l **7**, 439–451 (1987)
21. Precupanu, A., Precupanu, T., Turinici, M., Apreutesei Dumitriu, N., Stamate, C., Satco, B.R., Văideanu, C., Apreutesei, G., Rusu, D., Gavriluţ, A.C., Apetrii, M.: Modern Directions in Multivalued Analysis and Optimization Theory (in Romanian). Venus Publishing House, Iaşi (2006)
22. Precupanu, A., Gavriluţ, A.: Set-valued Lusin type theorem for null-null-additive set multifunctions, Fuzzy Sets Syst. **204**, 106–116 (2012)
23. Sharma, P., Nagar, A.: Topological dynamics on hyperspaces. Appl. Gen. Topol. **11**(1), 1–19 (2010)
24. Solecki, S.: G_δ ideals of compact sets. J. Eur. Math. Soc. **13**, 853–882 (2011)
25. Wicks, K.R.: Fractals and Hyperspaces. Springer, Berlin (1991)
26. Wijsman, R.: Convergence of sequences of convex sets, cones and functions. II. Trans. Amer. Math. Soc. **123**(1), 32–45 (1966)

Chapter 2
A mathematical-physical approach on regularity in hit-and-miss hypertopologies for fuzzy set multifunctions

2.1 Introduction

Hausdorff, Vietoris, Wijsman, Fell, Attouch-Wets and other well-known hypertopologies have been intensively studied in the last decades due to their various applications in Optimization, Convex Analysis, Economics, Image Processing, Sound Analysis and Synthesis (Beer [6], Apreutesei [3], Hu and Papageorgiou [20] concerning Vietoris topology). Results involving the Hausdorff distance were obtained by Lorenzo and Maio [8] in melodic similarity, Lu et al. [26]—an approach to word image matching, etc. Recently, it was shown that using proximity, all hypertopologies known so far are of the type hit-and-miss, which led to the unification of all hypertopologies under one topology called *the Bombay Hypertopology* [9].

On the other hand, fractals are multiscale objects, which seem to describe phenomena behaviour better than one does using traditional mathematical models. In consequence, fractal-based techniques lie at the heart of many areas, such as pure mathematics, applied mathematics, physics, etc. Kunze et al. [21] and Wicks [33] developed hyperspace theories concerning the Hausdorff metric and the Vietoris topology, as a foundation for self-similarity and fractality. In fact, for many years, topological methods were used in many fields to study the chaotic nature in dynamical systems (Sharma and Nagar [31], Wang et al. [32], Goméz-Rueda et al. [19], Li [23], Liu et al. [25], Ma et al. [27], Fu and Xing [12], etc.). These phenomena seem to be collective (set-valued), emerging out of many segregated components, having collective dynamics of many units of individual systems. This arised the need of a topological study of such collective dynamics. Recent studies of dynamical systems, in engineering and physical sciences, have revealed that the underlying dynamics is set-valued (collective), and not of a normal, individual kind, as it was usually studied before.

The reader can also refer to Lewin et al. [22] and Brown [7] for some approaches of topology in psychology. We also mention different aspects concerning generalized fractals in hyperspaces endowed with Hausdorff hypertopology, or, more

A. Gavriluţ et al., *Atomicity through Fractal Measure Theory*,
https://doi.org/10.1007/978-3-030-29593-6_2

generally, with Vietoris hipertopology (Andres and Fiser [1], Andres and Rypka [2], Banakh and Novosad [4], Kunze et al. [21]).

Since in some examples of fractals (like neural networks and the circulatory system), the uniform property of the Hausdorff topology is inappropriate, one could intend to choose a convenient topology on the set of values of the studied multifunctions. In this sense, Wijsman topology may be preferred instead of Hausdorff topology because Wijsman topology could describe better the pointwise properties of fractals.

On the other hand, recently, domain theory has been studied in theoretical computer science, as a mathematical theory of semantics of programming languages (Edalat [11], Gierz et al. [18] etc.). In this context, (hyper)topological notions from mathematical analysis as well as measure theory, dynamical systems or fractality can be considered via domain theory, obtaining computational models. Namely, in denotational semantics and domain theory, power domains are domains of nondeterministic and concurrent computations. Domain theory was introduced by Scott in theoretical computer science as a mathematical theory of semantics of programming languages.

Together with the increasing interest in hypertopologies, non-additive set multifunctions theories developed. In this context, regularity is known as an important continuity property with respect to different topologies, but, at the same time, it can be interpreted as an approximation property. Using regularity, we can approximate "unknown" sets by other sets for which we have more information. Usually, from a mathematical perspective, this approximation is done from the left by closed sets, or more restrictive, by compact sets and/or from the right by open sets. As a mathematical direct application of regularity, the classical Lusin's theorem concerning the existence of continuous restrictions of measurable functions is very important and useful for discussing different kinds of approximation of measurable functions defined on special topological spaces and for numerous applications in the study of convergence of sequences of Sugeno and Choquet integrable functions (see Li et al. [24] for an application of Lusin theorem), in the study of the approximation properties of neural networks, as the learning ability of a neural network is closely related to its approximating capabilities. Also, regular Borel measures are important tools in studies on the Kolmogorov fractal dimension (Barnsley [5], Mandelbrot [28]). Also, Lebesgue measure is a remarkable example of a regular measure.

Our unifying mathematical-physical point of view on fractality, hypertopologies and regularity was initiated in recent works, for instance [13–16] etc.

The present chapter represents a continuation of Chapter 1 in what concerns several main hypertopologies and their applications in regularity.

2.2 Hit-and-miss hypertopologies. An overview

Hausdorff, Vietoris and Wijsman hypertopologies are remarkable examples of the so-called hit-and-miss hypertopologies. Like some physical concepts, these hypertopologies, although composed of two independent parts, upper and lower hypertopologies, become consistent when they are considered together. For instance, in physical terms, the non-differentiability of the curve motion of the physical object involves the simultaneous definition at any point of the curve, of two differentials (left and right). Since we cannot favour one of the two differentials, the only solution is to consider them simultaneously through a complex differential. Its application, multiplied by dt, where t is an affine parameter, to the field of space coordinates implies complex speed fields.

We used the following (selected) references: Apreutesei [3], Beer [6], Gavriluţ and Apreutesei [17], Kunze et al. [21], Hu and Papageorgiou [20, Ch. 1], Precupanu et al. [29, Ch. 1], Apreutesei in Precupanu et al. [30, Ch. 8], Maio and Naimpally [9] etc.

We now briefly recall and list the definitions and main properties of the above-mentioned hypertopologies:

2.2.1 Vietoris topology

Let (X, τ) be a Hausdorff, topological space (we also introduced Vietoris topology in Chapter 1, but for Hausdorff linear topological spaces).

$M^- = \{C \in \mathcal{P}_0(X); M \cap C \neq \emptyset\}$ (i.e., *C hits M*), $M^+ = \{C \in \mathcal{P}_0(X); C \subseteq M\}$ (i.e., *C misses cM*), $S_{UV} = \{D^+; D \in \tau\}$ and $S_{LV} = \{D^-; D \in \tau\}$.

Vietoris topology $\widehat{\tau}_V$ on $\mathcal{P}_0(X)$ has as a subbase the class $S_{UV} \cup S_{LV}$ and it is the supremum $\widehat{\tau}_V = \widehat{\tau}_V^+ \cup \widehat{\tau}_V^-$ of the lower and upper Vietoris topologies:

$\widehat{\tau}_V^+$—*the upper Vietoris topology* ($\widehat{\tau}_V^-$—*the lower Vietoris topology*, respectively) is the topology which has as a subbase the class S_{UV} (S_{LV}, respectively).

For $U, V \in \tau$, define $\mathcal{B}_{U,V_1,V_2,\ldots,V_k} = U^+ \cap V_1^- \cap V_2^- \cap \ldots \cap V_k^-$.

The family $\mathcal{B}_{U,V_1,V_2,\ldots,V_k}$ of such subsets, where $U, V_1, V_2, \ldots, V_k \in \tau$, is a base for the topology $\widehat{\tau}_V$ and the family of subsets $\mathcal{B}_U = U^+$ ($\mathcal{B}_V = V^-$, respectively) is a base for $\widehat{\tau}_V^+$ ($\widehat{\tau}_V^-$, respectively).

In different continuity properties (regularity for instance), the following observation is used:

Remark 2.2.1 (Precupanu et al. [30, Ch. 8]) A net $(A_i)_{i \in I} \subset \mathcal{P}_0(X)$ is:

(i) $\widehat{\tau}_V^-$-convergent to $A_0 \in \mathcal{P}_0(X)$ if for every $V \in \tau$, with $A_0 \cap V \neq \emptyset$, $\exists i_V \in I$ so that for every $i \in I, i \geq i_V$, we have $A_i \cap V \neq \emptyset$;
(ii) $\widehat{\tau}_V^+$-convergent to $A_0 \in \mathcal{P}_0(X)$ if for every $V \in \tau$, with $A_0 \subset V, \exists i_V \in I$ so that for every $i \in I, i \geq i_V$, we have $A_i \subset V$.

2.2.2 Wijsman topology

In what follows, let (X, d) be a metric space.

For the convenience of the reader, we recall in what follows several considerations on Wijsman topology, that we have mentioned in Chapter 1.

Wijsman topology τ_W on $\mathcal{P}_0(X)$ is the supremum of the *upper Wijsman topology* τ_W^+ and the *lower Wijsman topology* τ_W^- : $\tau_W = \tau_W^+ \cup \tau_W^-$.

The family

$$\mathcal{F} = \{M \in \mathcal{P}_0(X); d(x, M) < \varepsilon\}_{\substack{x \in X \\ \varepsilon > 0}} \cup \{M \in \mathcal{P}_0(X); d(x, M) > \varepsilon\}_{\substack{x \in X \\ \varepsilon > 0}}$$

is a subbase for τ_W on $\mathcal{P}_0(X)$.

Let $M \in \mathcal{P}_0(X)$, $\{x_1, x_2, \ldots, x_n\} \subset X$, $\varepsilon > 0$ be arbitrarily chosen.

τ_W^- (τ_W^+, respectively) is generated by the family $U_W^-(M, x_1, x_2, \ldots, x_n, \varepsilon) = \{N \in \mathcal{P}_0(X); d(x_i, N) < d(x_i, M) + \varepsilon$, for every $i = \overline{1, n}\}$ ($U_W^+(M, x_1, x_2, \ldots, x_n, \varepsilon) = \{N \in \mathcal{P}_0(X); d(x_i, M) < d(x_i, N) + \varepsilon$, for every $i = \overline{1, n}\}$, respectively).

Proposition 2.2.2 (Apreutesei [3]; Ch. 8 in Precupanu et al. [30]) $\tau_W^- = \tau_V^-$.

Remark 2.2.3

(I) Suppose $\{M_i\}_{i \in I} \subset \mathcal{P}_0(X)$. The following statements are equivalent:

(i) $M_i \overset{\tau_W}{\to} M \in \mathcal{P}_0(X)$;

(ii) For every $x \in X$, $d(x, M_i) \overset{p}{\to} d(x, M)$ (pointwise convergence);

(iii) $M_i \overset{\tau_W^+}{\to} M$ and $M_i \overset{\tau_W^-}{\to} M$.

(II) (i) $M_i \overset{\tau_W^+}{\to} M$ if and only if for every $x \in X$, $\liminf_i d(x, M_i) \geq d(x, M)$ (i.e., for every $0 < \varepsilon < \varepsilon'$ with $S(x, \varepsilon') \cap M = \emptyset$, there is $i_0 \in I$ so that for every $i \in I$, with $i \geq i_0$, we have $S(x, \varepsilon) \cap M_i = \emptyset$);

(ii) $M_i \overset{\tau_W^-}{\to} M$ if and only if for every $x \in X$, $\limsup_i d(x, M_i) \leq d(x, M)$ (i.e., for every $D \in \tau_d$ with $D \cap M \neq \emptyset$, there is $i_0 \in I$ so that for every $i \in I$, with $i \geq i_0$ we have $D \cap M_i \neq \emptyset$).

Remark 2.2.4 ([17])

(i) If (X, d) is a complete, separable metric space, then $\mathcal{P}_f(X)$ with the Wijsman topology is a Polish space (Beer [6]). Moreover, the space $(\mathcal{P}_f(X), \tau_W)$ is Polish if and only if (X, d) is Polish.

(ii) $(\mathcal{P}_f(X), \tau_W)$ is a Tychonoff space.

(X, d) is separable if and only if $\mathcal{P}_f(X)$ is either metrizable, first-countable or second-countable. The dependence of the Wijsman topology on the metric d is quite strong. Even if two metrics are uniformly equivalent, they may generate different Wijsman topologies. Necessary and sufficient conditions for two metrics to induce the same Wijsman topology have been found.

2.2.3 Hausdorff topology

In recent years, due to the development of computational graphics (for instance, in the automatic recognition of figures problems), it became a necessity to measure accurately the matching, i.e., to calculate the distance between two sets of points. This led to the need to operate with an acceptable distance, which has to satisfy the first condition in the definition of a distance: the distance is zero if and only if the overlap is perfect. An appropriate metric in these issues is the Hausdorff metric on which we will refer in the following and which, roughly speaking, measures the degree of overlap of two compact sets.

For instance, the Cantor set $\mathcal{C} \in \mathcal{P}_k(\mathbb{R})$ and its "steps" $I_n \in \mathcal{P}_k(\mathbb{R}), \forall n \in \mathbb{N}$ (Kunze et al. [21]).

Let be $M, N \in \mathcal{P}_f(X)$. The *Hausdorff-Pompeiu pseudometric* h on $\mathcal{P}_f(X)$ is the "greatest" of all distances from any point in one of these two sets, to the nearest point from the other set. It is defined by

$$(*)\ h(M,N) = \max\{e(M,N), e(N,M)\} \text{ (see Chapter 1).}$$

The topology induced by the Hausdorff pseudometric h is called the *Hausdorff hipertopology* τ_H *on* $\mathcal{P}_f(X)$.

On $\mathcal{P}_{bf}(X)$, h becomes a veritable metric. If, in addition, X is complete, then the same is $\mathcal{P}_f(X)$ (Hu and Papageorgiou [20]).

We observe that $e(M,N) \leq e(M,P)$, for every $M, N, P \in \mathcal{P}_f(X)$, with $P \subseteq N$ and $e(M,P) \leq e(N,P)$, for every $M, N, P \in \mathcal{P}_f(X)$, with $M \subseteq N$.

Generally, even if $M, N \in \mathcal{P}_k(X)$, then $e(M,N) \neq e(N,M)$.

If $M \in \mathcal{P}_f(X)$ and $\varepsilon > 0$ is arbitrary, but fixed, we consider the *ε-dilatation of the set M*

$$S(M,\varepsilon) = \{x \in X; \exists m \in M, d(x,m) < \varepsilon\}$$

$$\left(= \bigcup_{m \in M} \{x \in X; m \in M, d(x,m) < \varepsilon\}\right).$$

Obviously, $M \subseteq S(M,\varepsilon)$.

Since $h(M,N) < \varepsilon$ iff $M \subset S(N,\varepsilon)$ and $N \subset S(M,\varepsilon)$, we have the following equivalent expression for $h(M,N)$:

$$(**)\ h(M,N) = \inf\{\varepsilon > 0; M \subset S(N,\varepsilon), N \subset S(M,\varepsilon)\}$$

($h(M,N)$ is the "smallest" $\varepsilon > 0$ which permits the ε-dilatation of M to cover N and the ε-dilatation of N to cover M).

In other words, $\tau_H = \tau_H^+ \cup \tau_H^-$, where τ_H^+ (*upper Hausdorff topology*), respectively, τ_H^- (*lower Hausdorff topology*) has as a base, the family $\{U^+(M,\varepsilon)\}_{\varepsilon>0}$, where $U^+(M,\varepsilon) = \{N \in \mathcal{P}_f(X); N \subset S(M,\varepsilon)\}$, respectively, the family $\{U^-(M,\varepsilon)\}_{\varepsilon>0}$, where $U^-(M,\varepsilon) = \{N \in \mathcal{P}_f(X); M \subset S(N,\varepsilon)\}$.

Another equivalent expression of the Hausdorff distance between two sets $M, N \in \mathcal{P}_f(X)$ is:

$$(***)\ h(M, N) = \sup\{|d(x, N) - d(x, M)|; x \in X\}.$$

and this highlights the uniform aspect of the Hausdorff topology: it is the topology on $\mathcal{P}_f(X)$ of uniform convergence on X of the distance functionals $x \mapsto d(x, M)$, with $M \in \mathcal{P}_f(X)$.

Hausdorff topology is invariant with respect to uniformly equivalent metrics (Apreutesei [3]).

Besides the properties of e and h listed in Section 1.2, we also mention the following:

Proposition 2.2.5

(I)

$$h(M_1 \cup M_2, N_1 \cup N_2) \le \max\{h(M_1, N_1), h(M_2, N_2)\},$$
$$\forall M_1, M_2, N_1, N_2 \in \mathcal{P}_f(X);$$

(II) If X is a Banach space, then:

(i)

$$h(\alpha M, \alpha N) = |\alpha| h(M, N), \forall \alpha \in \mathbb{R}, \forall M, N \in \mathcal{P}_f(X);$$

(ii)

$$h(M + P, N + P) \le h(M, N), \forall M, N, P \in \mathcal{P}_f(X);$$

Moreover

(ii′)

$$h(\sum_{i=1}^{p} M_i, \sum_{i=1}^{p} N_i) \le \sum_{i=1}^{p} h(M_i, N_i), \forall M_i, N_i \in \mathcal{P}_f(X), i = \overline{1, p}.$$

If, particularly, $X = \mathbb{R}$, and $a, b, c, d \in \mathbb{R}$, with $a < b, c < d$, then

$$h([a, b], [c, d]) = \max\{|a - c|, |b - d|\}.$$

Remark 2.2.6 (Kunze et al. [21]) The Hausdorff metric has some interesting characteristics:

(i) It is possible for a sequence of finite sets to converge to an uncountable set:

$\forall n \geq 1$, let be $M_n = \{0, \frac{1}{n}, \frac{2}{n}, \ldots, \frac{n-1}{n}, 1\} (\subset [0, 1])$ (all of them are finite sets). Since $h(M_n, [0, 1]) = \frac{1}{2n}$, then $M_n \xrightarrow{h} [0, 1]$ (in $\mathcal{P}_k(\mathbb{R})$), but $[0, 1]$ is uncountable.

(ii) Adding or removing a single point often influences the Hausdorff distance between two (compact) sets: if $M = [0, 1]$ and $N = [0, 1] \cup \{x\}$, where $x \notin [0, 1]$, then $h(M, N) = \max\{-x, x - 1\}$ (so, it is a function of x).

(iii) $I_n \xrightarrow{h} \mathcal{C}$ and $J_n \xrightarrow{h} \mathcal{C}$ where $(I_n)_{n\in\mathbb{N}}$ (respectively, $(J_n)_{n\in\mathbb{N}}$) are the steps in the construction of Cantor set $\mathcal{C}$.

Remark 2.2.7 Concerning the relationships among Hausdorff, Wijsman and Vietoris topologies, the reader can refer to Chapter 1 and also to [3, 17] [29, Ch. 1] [30, Ch. 8].

Remark 2.2.8 Hausdorff metric on $\mathcal{P}_k(X)$ is an essential tool in the study of fractals and their generalizations: hyperfractals, multifractals and superfractals (see [1, 2, 21]).

Barnsley [5] calls the space $(\mathcal{P}_k(X), h)$, *the life space of fractal.* Recently, Banakh and Novosad [4] proposed a fractal approach using Vietoris topology (in a more general setting than the one used for the Hausdorff topology).

2.3 Regular set multifunctions

Suppose that T is a locally compact, Hausdorff space, $\mathcal{C}$ a ring of subsets of T and X a real normed space. Usually, it is assumed that $\mathcal{C}$ is $\mathcal{B}_0$ ($\mathcal{B}'_0$, respectively)—the Baire δ-ring (σ-ring, respectively) generated by compact sets, which are G_δ (i.e., countable intersections of open sets) or $\mathcal{C}$ is $\mathcal{B}$ ($\mathcal{B}'$, respectively)—the Borel δ-ring (σ-ring, respectively) generated by the compact sets of T.

$\mathcal{B}_0 \subset \mathcal{B} \subset \mathcal{B}'$, $\mathcal{B}_0 \subset \mathcal{B}'_0$. If T is metrizable or if it has a countable base, then any compact set $K \subset T$ is G_δ. In this case $\mathcal{B}_0 = \mathcal{B}$ (Dinculeanu [10, Ch. III, p. 187]) so $\mathcal{B}'_0 = \mathcal{B}'$.

By $\mathcal{K}$ we denote the family of all compact subsets of T and by $\mathcal{D}$ the family of all open subsets of T.

Regularity can be considered as a property of continuity with respect to a topology on $\mathcal{P}(T)$, the family of all subsets of T (Dinculeanu [10, Ch. III, p. 197]):

For every $K \in \mathcal{K}$ and every $D \in \mathcal{D}$, with $K \subset D$, we denote $\mathcal{I}(K, D) = \{A \subset T; K \subset A \subset D\}$.

Since $\mathcal{I}(K, D) \cap \mathcal{I}(K', D') = \mathcal{I}(K \cup K', D \cap D')$, for every $\mathcal{I}(K, D), \mathcal{I}(K', D')$, the family $\{\mathcal{I}(K, D)\}_{\substack{K\in\mathcal{K}\\ D\in\mathcal{D}}}$ is a base of a topology $\tilde{\tau}$ on $\mathcal{P}(T)$. $\tilde{\tau}$ also denotes the topology induced on any subfamily $\mathcal{S} \subset \mathcal{P}(T)$ of subsets of T.

By $\tilde{\tau}_l$ ($\tilde{\tau}_r$, respectively) we denote the topology induced on $\{\mathcal{I}(K)\}_{K\in\mathcal{K}} = \{\{A \subset T; K \subset A\}\}_{K\in\mathcal{K}}$ ($\{\mathcal{I}(D)\}_{D\in\mathcal{D}} = \{\{A \subset T; A \subset D\}\}_{D\in\mathcal{D}}$, respectively) (Dinculeanu [10, Ch. III, pp. 197–198]).

Definition 2.3.1 A class $\mathcal{F} \subset \mathcal{P}(T)$ is *dense* in $\mathcal{P}(T)$ with respect to the topology induced by $\tilde{\tau}$ if for every $K \in \mathcal{K}$ and every $D \in \mathcal{D}$, with $K \subset D$, there is $A \in \mathcal{C}$ such that $K \subset A \subset D$.

Since T is locally compact, the following statements can be easily verified (Dinculeanu [10, Ch. III, p. 197]):

Remark 2.3.2

(1) $\mathcal{B}_0, \mathcal{B}, \mathcal{B}'_0, \mathcal{B}'$ are dense in $\mathcal{P}(T)$ with respect to the topology induced by $\tilde{\tau}$.
(2) (i) For every $A \in \mathcal{C}$, there exists $D \in \mathcal{D} \cap \mathcal{C}$ so that $A \subset D$.
(ii) If $\mathcal{C}$ is $\mathcal{B}$ or $\mathcal{B}'$, then for every $A \in \mathcal{C}$, there exist $K \in \mathcal{K} \cap \mathcal{C}$ and $D \in \mathcal{D} \cap \mathcal{C}$ so that $K \subset A \subset D$.

Let $\mu : \mathcal{C} \to \mathcal{P}_0(X)$ be an arbitrary set multifunction.

Definition 2.3.3 μ is said to be *monotone* or *fuzzy or monotone-continuous* (with respect to the inclusion of sets) if $\mu(A) \subseteq \mu(B)$, for every $A, B \in \mathcal{C}$ with $A \subseteq B$.

Example 2.3.4 (of Monotone Set Multifunctions)

(i) Let $\mathcal{C}$ be a ring of subsets of an abstract space T, $m : \mathcal{C} \to \mathbb{R}_+$ a finitely additive set function and $\mu : \mathcal{C} \to \mathcal{P}_{bf}(\mathbb{R})$ the set multifunction defined for every $A \in \mathcal{C}$ by:

$$\mu(A) = \begin{cases} [-m(A), m(A)], & \text{if } m(A) \leq 1 \\ \\ [-m(A), 1], & \text{if } m(A) > 1 \end{cases}.$$

We easily observe that μ is monotone on $\mathcal{C}$.

(ii) Let $\nu_1, \ldots, \nu_p : \mathcal{C} \to \mathbb{R}_+$, be p finitely additive set functions, where $\mathcal{C}$ is a ring of subsets of an abstract space T. We consider the set multifunction $\mu : \mathcal{C} \to \mathcal{P}_f(\mathbb{R})$, defined for every $A \in \mathcal{C}$ by

$$\mu(A) = \{\nu_1(A), \nu_2(A), \ldots, \nu_p(A)\}.$$

Then the set multifunction $\mu^{\vee} : \mathcal{C} \to \mathcal{P}_f(\mathbb{R})$, defined for every $A \in \mathcal{C}$ by

$$\mu^{\vee}(A) = \overline{\bigcup_{\substack{B\subset A,\\ B\in\mathcal{C}}} \mu(B)}.$$

is monotone.

In what follows, let $\mu : (\mathcal{C}, \tau_1) \to (\mathcal{P}_f(X), \tau_2)$ be a monotone set multifunction, where $\tau_1 \in \{\tilde{\tau}, \tilde{\tau}_l, \tilde{\tau}_r\}$ and $\tau_2 \in \{\tau_H, \tau_W, \tau_V\}$.

Let also be $\mathcal{B}_1 \in \{\{\mathcal{I}(K, D)\}_{\substack{K\in\mathcal{K}\\ D\in\mathcal{D}}}, \{\mathcal{I}(K)\}_{K\in\mathcal{K}}, \{\mathcal{I}(D)\}_{D\in\mathcal{D}}\}$, respectively, $\mathcal{B}_2$, bases for τ_1, respectively τ_2 (as discussed in Section 2.2).

$\tau_2 = \tau_2^+ \cup \tau_2^-$, where $\tau_2^+ \in \{\tau_H^+, \tau_V^+, \tau_W^+\}$ and $\tau_2^- \in \{\tau_H^-, \tau_V^-, \tau_W^-\}$.

In a unifying way,

Definition 2.3.5 A set $A \in \mathcal{C}$ is said to be (τ_2-)*regular* if $\mu : (\mathcal{C}, \tau_1) \to (\mathcal{P}_f(X), \tau_2)$ is continuous at A, that is, for every $\mathcal{V} \in \mathcal{B}_2$, with $\mu(A) \in \mathcal{V}$, there exists $\widetilde{\mathcal{V}} \in \mathcal{B}_1$ so that $\mu(\widetilde{\mathcal{V}} \cap \mathcal{C}) \subset \mathcal{V}$ (or, equivalently, for every $(A_i)_{i\in I}, A \subset \mathcal{C}$, with $A_i \overset{\tau_1}{\to} A$, it results $\mu(A_i) \overset{\tau_2}{\to} \mu(A)$).

When τ_1 is $\tilde{\tau}$, or $\tilde{\tau}_l$ or $\tilde{\tau}_r$, respectively, we get the notions of (τ_2-)*regularity*, (τ_2-)R_l*-regularity* (inner regularity) or (τ_2-)R_r*-regularity* (outer regularity).

Precisely, we have:

Proposition 2.3.6 *A is:*

(i) regular iff for every $\mathcal{V} \in \mathcal{B}_2$, with $\mu(A) \in \mathcal{V}$, there exist $K \in \mathcal{K} \cap \mathcal{C}$, $K \subset A$ and $D \in \mathcal{D} \cap \mathcal{C}$, $D \supset A$ so that for every $B \in \mathcal{C}$, with $K \subset B \subset D$, we have $\mu(B) \in \mathcal{V}$;

(ii) R_l-regular iff for every $\mathcal{V} \in \mathcal{B}_2$, with $\mu(A) \in \mathcal{V}$, there exists $K \in \mathcal{K}\cap\mathcal{C}$, $K \subset A$ so that for every $B \in \mathcal{C}$, with $K \subset B \subset A$, we have $\mu(B) \in \mathcal{V}$;

(iii) R_r-regular iff for every $\mathcal{V} \in \mathcal{B}_2$, with $\mu(A) \in \mathcal{V}$, there exists $D \in \mathcal{D}\cap\mathcal{C}$, $D \supset A$ so that for every $B \in \mathcal{C}$, with $A \subset B \subset D$, we have $\mu(B) \in \mathcal{V}$.

Remark 2.3.7 Every $K \in \mathcal{K}$ is R_l-regular and every $D \in \mathcal{D}$ is R_r-regular.

The following results can be proved using the above definitions:

Proposition 2.3.8

(i) A set A is (τ_2)-regular if and only if it is (τ_2-)R_l-regular and (τ_2-)R_r-regular.

(ii) A is (τ_2-)regular (R_l-regular or R_r-regular, respectively) if and only if it is (τ_2^+-) and (τ_2^--)regular (R_l-regular or R_r-regular, respectively).

Theorem 2.3.9 *Suppose $\mu_1, \mu_2 : \mathcal{C} \to (\mathcal{P}_f(X), \tau_2)$ are two monotone set multifunctions.*

(i) *If μ_1, μ_2 are R_l-regular, then $\mu_1 = \mu_2$ on $\mathcal{C}$ if and only if $\mu_1 = \mu_2$ on $\mathcal{K} \cap \mathcal{C}$;*

(ii) *If μ_1, μ_2 are R_r-regular, then $\mu_1 = \mu_2$ on $\mathcal{C}$ if and only if $\mu_1 = \mu_2$ on $\mathcal{D} \cap \mathcal{C}$.*

Remark 2.3.10 For $\tau_2 = \tau_H, \tau_W$ or τ_V, respectively, we particularly get the notions of regularity as we defined and studied in [13, 14, 17]. For instance, if $\tau_2 = \tau_H$, then, by its monotonicity, μ is (in the sense of [13]):

(i) *regular* if for every $\varepsilon > 0$, there are $K \in \mathcal{K}\cap\mathcal{C}$, $K \subset A$ and $D \in \mathcal{D}\cap\mathcal{C}$, $D \supset A$ so that $h(\mu(A), \mu(B)) < \varepsilon$, for every $B \in \mathcal{C}$, with $K \subset B \subset D$.

(ii) R_l*-regular* if for every $\varepsilon > 0$, there exists $K \in \mathcal{K} \cap \mathcal{C}, K \subset A$ so that $h(\mu(A), \mu(B)) = e(\mu(A), \mu(B)) < \varepsilon$, for every $B \in \mathcal{C}$, with $K \subset B \subset A$.

(iii) R_r*-regular* if for every $\varepsilon > 0$, there exists $D \in \mathcal{D} \cap \mathcal{C}, D \supset A$ such that $h(\mu(A), \mu(B)) = e(\mu(B), \mu(A)) < \varepsilon$, for every $B \in \mathcal{C}$, with $A \subset B \subset D$.

In fact, one may easily observe that (with respect to τ_H):

(i) μ is regular iff for every $\varepsilon > 0$, there are $K \in \mathcal{K} \cap \mathcal{C}, K \subset A$ and $D \in \mathcal{D} \cap \mathcal{C}, D \supset A$ so that $e(\mu(D), \mu(K)) < \varepsilon$;
(ii) μ is R_l-regular iff for every $\varepsilon > 0$, there is $K \in \mathcal{K} \cap \mathcal{C}, K \subset A$ so that $e(\mu(A), \mu(K)) < \varepsilon$;
(iii) μ is R_r-regular iff for every $\varepsilon > 0$, there is $D \in \mathcal{D} \cap \mathcal{C}, D \supset A$ so that $e(\mu(D), \mu(A)) < \varepsilon$,

that is, in each case, we find an alternative expression of regularity as an approximation property.

References

1. Andres, J., Fiser, J.: Metric and topological multivalued fractals. Int. J. Bifur. Chaos Appl. Sci. Eng. **14**(4), 1277–1289 (2004)
2. Andres, J., Rypka, M.: Multivalued fractals and hyperfractals. Int. J. Bifur. Chaos Appl. Sci. Eng. **22**(1), 1250009, 27 pp. (2012)
3. Apreutesei, G.: Families of subsets and the coincidence of hypertopologies. Ann. Alexandru Ioan Cuza Univ. Math. **XLIX**, 1–18 (2003)
4. Banakh, T., Novosad, N.: Micro and macro fractals generated by multi-valued dynamical systems, arXiv: 1304.7529v1 [math.GN] (2013)
5. Barnsley, M.: Fractals Everywhere. Academic, Boston (1988)
6. Beer, G.: Topologies on Closed and Closed Convex Sets. Kluwer Academic Publishers, Dordrecht (1993)
7. Brown, S.: Memory and mathesis: for a topological approach to psychology. Theory Cult. Soc. **29**(4–5), 137–164 (2012)
8. di Lorenzo, P., di Maio, G.: The Hausdorff metric in the melody space: a new approach to melodic similarity. In: The 9th International Conference on Music Perception and Cognition, Alma Mater Studiorum University of Bologna, August 22–26 (2006)
9. di Maio, G., Naimpally, S.: Hit-and-far-miss hypertopologies. Mat. Vesnik **60**, 59–78 (2008)
10. Dinculeanu, N.: Measure Theory and Real Functions (in Romanian). Editura Didactică şi Pedagogică, Bucureşti (1964)
11. Edalat, A.: Dynamical systems, measures and fractals via domain theory. Inf. Comput. **120**(1), 32–48 (1995)
12. Fu, H., Xing, Z.: Mixing properties of set-valued maps on hyperspaces via Furstenberg families. Chaos Solitons Fractals **45**(4), 439–443 (2012)
13. Gavriluţ, A.: Regularity and autocontinuity of set multifunctions. Fuzzy Sets Syst. **161**, 681–693 (2010)
14. Gavriluţ, A.: Continuity properties and Alexandroff theorem in Vietoris topology. Fuzzy Sets Syst. **194**, 76–89 (2012)
15. Gavriluţ, A., Agop, M.: A Mathematical Approach in the Study of the Dynamics of Complex Systems. Ars Longa Publishing House, Iaşi (2013). (in Romanian)

16. Gavriluţ, A., Agop, M.: Approximation theorems for set multifunctions in Vietoris topology. Physical implications of regularity. Iran. J. Fuzzy Syst. **12**(1), 27–42 (2015)
17. Gavriluţ, A., Apreutesei, G.: Regularity aspects of non-additive set multifunctions. Fuzzy Sets Syst. (2016). https://doi.org/10.1016/j.fss.2016.02.003
18. Gierz, G., Hofmann, K.H., Keimel, K., Lawson, J.D., Mislove, M.W., Scott, D.S.: Continuous Lattices and Domains. Encyclopedia of Mathematics and its Applications, vol. 93. Cambridge University Press, Cambridge (2003)
19. Gómez-Rueda, J.L., Illanes, A., Méndez, H.: Dynamic properties for the induced maps in the symmetric products. Chaos Solitons Fractals **45**(9–10), 1180–1187 (2012)
20. Hu, S., Papageorgiou, N.S.: Handbook of Multivalued Analysis, vol. I. Kluwer Academic Publishers, Dordrecht (1997)
21. Kunze, H., La Torre, D., Mendivil, F., Vrscay, E.R.: Fractal Based Methods in Analysis. Springer, New York (2012)
22. Lewin, K., Heider, G.M., Heider, F.: Principles of Topological Psychology. McGraw-Hill, New York (1936)
23. Li, R.: A note on stronger forms of sensitivity for dynamical systems. Chaos Solitons Fractals **45**(6), 753–758 (2012)
24. Li, J., Li, J., Yasuda, M.: Approximation of fuzzy neural networks by using Lusin's theorem, Kioto University Research Information Repository, pp. 86–92 (2007)
25. Liu, L., Wang, Y., Wei, G.: Topological entropy of continuous functions on topological spaces. Chaos Solitons Fractals **39**(1), 417–427 (2009)
26. Lu, Y., Tan, C.L., Huang, W., Fan, L.: An approach to word image matching based on weighted Hausdorff distance. In: Proceedings of International Conference on Document Analysis and Recognition, pp. 921–925 (2001)
27. Ma, X., Hou, B., Liao, G.: Chaos in hyperspace system. Chaos Solitons Fractals **40**(2), 653–660 (2009)
28. Mandelbrot, B.B.: The Fractal Geometry of Nature. W.H. Freiman, New York (1983)
29. Precupanu, A., Gavriluţ, A.: Set-valued Lusin type theorem for null-null-additive set multifunctions. Fuzzy Sets Syst. **204**, 106–116 (2012)
30. Precupanu, A., Precupanu, T., Turinici, M., Apreutesei Dumitriu, N., Stamate, C., Satco, B.R., Văideanu, C., Apreutesei, G., Rusu, D., Gavriluţ, A.C., Apetrii, M.: Modern Directions in Multivalued Analysis and Optimization Theory. Venus Publishing House, Iaşi (2006). (in Romanian)
31. Sharma, P., Nagar, A.: Topological dynamics on hyperspaces. Appl. Gen. Topol. **11**(1), 1–19 (2010)
32. Wang, Y., Wei, G., Campbell, W.H., Bourquin, S.: A framework of induced hyperspace dynamical systems equipped with the hit-or-miss topology. Chaos Solitons Fractals **41**(4), 1708–1717 (2009)
33. Wicks, K.R.: Fractals and Hyperspaces. Springer, Berlin (1991)

Chapter 3
Non-atomic set multifunctions

3.1 Basic notions, terminology and results

The study of non-additive set functions and set multifunctions has recently received a special attention, because of its applications in statistics, sociology, biology, theory of games or economic mathematics. For example, Choquet [2] has been conducted to this subject by his research in the potential theory (the non-additive measures being called, in certain conditions, *capacities*). The results of Choquet led to the development of some theories with important applications in statistics, games theory and artificial intelligence. Aumann and Shapley [1], Denneberg [4], Dinculeanu [5], Dobrakov [6], Drewnowski [7], Klein and Thompson [16], Lipecki [18], Sugeno [25], Precupanu [22], Pap [19, 20], Suzuki [26], Wu and Bo [27], Gavriluţ [8–11], Gavriluţ and Croitoru [12] and many other authors investigated these fields of additive or non-additive set functions and multivalued set functions.

In the probability theory field, a generalized theory (called *the evidence theory*) based on two types of non-additive measures was initiated by Dempster [3] and Shafer [24].

Fuzzy measures, an important part of classical Measure Theory, were introduced and studied by Sugeno [25] in 1974. Since then, considerable applications have been found in physics, biology, medicine (Pham, Brandl, Nguyen and Nguyen [21] in prediction of osteoporotic fractures), theory of probabilities, economic mathematics, human decision making (Liginlal and Ow [17]). Certain results of classical Measure Theory (such as regularity, extensions, decompositions, integral) were transferred in the fuzzy case. For example, Kawabe [15] considered fuzzy measures taking values in a Riesz space; Zhang and Wang [28], Zhang and Guo [29] defined and studied a set-valued fuzzy integral, which is a natural extension of single-valued fuzzy integral; Guo and Zhang [13] introduced and studied the notion of fuzzy multivalued measure taking values in $\mathcal{P}_0(\overline{\mathbb{R}}_+)$, the family of nonempty subsets of $\overline{\mathbb{R}}_+$.

In mathematical economics, a special attention has been given to the notion of an atom of a measure. The traditional concept of an economy, where no individual

A. Gavriluţ et al., *Atomicity through Fractal Measure Theory*,
https://doi.org/10.1007/978-3-030-29593-6_3

agent can influence the outcome of a collective activity is described by an atomless measure (or non-atomic measure) on the space of agents characterizations, i.e., a measure ν on $\mathcal{A}$ so that $\nu(\{a\}) = 0$, for every $A \in \mathcal{A}$. The concept of an atomless measure space of economic agents is due to Aumann [1].

In this chapter, we consider the notion of a fuzzy multivalued set function (or set multifunction) (introduced by Guo and Zhang [13]) in the case of multivalued set functions taking values in $\mathcal{P}_f(X)$, the family of nonempty closed subsets of a real normed space X. Also, we introduce the concepts of atom and non-atomicity of a multivalued set function and present some of their properties. We also establish an extension result by preserving non-atomicity for a fuzzy multisubmeasure/multimeasure. Several special considerations will be given to $\mathcal{P}_{bf}(X)$-valued set multifunctions, X being a Banach space, with respect to the Hausdorff topology. The motivation of our study is given to the fact that non-atomic additive (or non-additive) measures have received a special attention because of the applications, in non-atomic games theory, for instance.

We now fix several notions used throughout this chapter.

Let T be an abstract nonvoid set and $\mathcal{C}$ a ring of subsets of T.

Definition 3.1.1 A set function $\nu : \mathcal{C} \to \overline{\mathbb{R}}_+$, with $\nu(\emptyset) = 0$, is said to be:

(i) *order continuous* (shortly, *o-continuous*) if $\lim\limits_{n\to\infty} \nu(A_n) = 0$, for every decreasing sequence of sets $(A_n)_{n\in\mathbb{N}^*} \subset \mathcal{C}$ (i.e., $A_n \supset A_{n+1}$, for every $n \in \mathbb{N}^*$) with $\bigcap\limits_{n=1}^{\infty} A_n = \emptyset$ (denoted by $A_n \searrow \emptyset$);
(ii) a *Dobrakov submeasure* if ν is a submeasure in the sense of Drewnowski [7] and it is also o-continuous.

Definition 3.1.2 Let $\nu : \mathcal{C} \to \mathbb{R}_+$ be a set function, with $\nu(\emptyset) = 0$.

(i) A set $A \in \mathcal{C}$ is said to be an *atom* of ν if $\nu(A) > 0$ and for every $B \in \mathcal{C}$, with $B \subset A$, we have $\nu(B) = 0$ or $\nu(A \backslash B) = 0$.
(ii) ν is said to be *non-atomic* if for each $A \in \mathcal{C}$ with $\nu(A) > 0$, there exists $B \in \mathcal{C}$, $B \subset A$, such that $\nu(B) > 0$ and $\nu(A \backslash B) > 0$.

Remark 3.1.3 Concerning the notions of atom and variation of a set function, the reader should consult the results about atomic non-additive set functions in Aumann and Shapley's famous book [1].

In the sequel, $(X, \|\cdot\|)$ will be a real normed space (having origin 0), with the distance d induced by its norm. We denote $cM = X \backslash M$, for every $M \in \mathcal{P}(X)$.

On $\mathcal{P}_0(X)$ we consider the Minkowski addition "$\overset{\bullet}{+}$", defined by:

$$M \overset{\bullet}{+} N = \overline{M + N}, \text{ for every } M, N \in \mathcal{P}_0(X).$$

For every $M, N \in \mathcal{P}_0(X)$, we denote

$$h(M, N) = \max\{e(M, N), e(N, M)\},$$

where e has the meaning from Chapter 1. We recall that h becomes an extended metric on $\mathcal{P}_f(X)$ (i.e., it is a metric which can also take the value $+\infty$) and h becomes a metric on $\mathcal{P}_{bf}(X)$. If X is a Banach space, then $(\mathcal{P}_{bf}(X), h)$ is a complete metric space (Hu and Papageorgiou [14]).

Definition 3.1.4 If $\mu : \mathcal{C} \to \mathcal{P}_0(X)$ is a multivalued set function (also called a set multifunction), then μ is said to be:

(i) a *multisubmeasure* (Gavriluţ [8]) if it is monotone, $\mu(\emptyset) = \{0\}$ and $\mu(A \cup B) \subseteq \mu(A) + \mu(B)$, for every $A, B \in \mathcal{C}$, with $A \cap B = \emptyset$ (or, equivalently, for every $A, B \in \mathcal{C}$);
(ii) a *multimeasure* if $\mu(\emptyset) = \{0\}$ and $\mu(A \cup B) = \mu(A) + \mu(B)$, for every $A, B \in \mathcal{C}$, with $A \cap B = \emptyset$.

Remark 3.1.5

(I) If μ is single-valued, with $\mu(\emptyset) = \{0\}$, then the monotonicity of μ implies that $\mu(A) = \{0\}$, for all $A \in \mathcal{C}$. Therefore, monotonicity finds a meaning in the case when the multivalued set function is not single-valued.
(II) If μ is $\mathcal{P}_f(X)$-valued, then in the Definition 3.1.4(ii) of a multimeasure, it usually appears "$\overset{\bullet}{+}$" instead of "$+$", because the sum of two closed sets is not always closed.
(III) Sometimes we shall assume μ to be $\mathcal{P}_f(X)$-valued, when we need h to be an extended metric or we shall deal with set multifunctions taking values in $\mathcal{P}_{bf}(X)$, when we use the Banach space structure of $\mathcal{P}_{bf}(X)$.

Definition 3.1.6 A set multifunction $\mu : \mathcal{C} \to \mathcal{P}_f(X)$, with $\mu(\emptyset) = \{0\}$, is said to be:

(i) *exhaustive* (with respect to h) if $\lim_{n\to\infty} |\mu(A_n)| = 0$, for every sequence of pairwise disjoint sets $(A_n)_{n\in\mathbb{N}^*} \subseteq \mathcal{C}$.
(ii) *order continuous* (shortly, *o-continuous*) (with respect to h) if $\lim_{n\to\infty} |\mu(A_n)| = 0$, for every $(A_n)_{n\in\mathbb{N}^*} \subseteq \mathcal{C}$, with $A_n \searrow \emptyset$.

Example 3.1.7 Let $\nu : \mathcal{C} \to \mathbb{R}_+$ and $\mu : \mathcal{C} \to \mathcal{P}_f(\mathbb{R})$, defined by $\mu(A) = [0, \nu(A)]$, for every $A \in \mathcal{C}$. If ν is a submeasure (finitely additive, respectively), then μ is a multisubmeasure (monotone multimeasure, respectively).

μ is called *the multisubmeasure* (*multimeasure*, respectively) *induced by* ν.

Definition 3.1.8 For a multivalued set function $\mu : \mathcal{C} \to \mathcal{P}_0(X)$, we consider the following set functions:

(i) $\overline{\mu} : \mathcal{P}(T) \to \overline{\mathbb{R}}_+$, called *the variation of* μ, defined for every $A \in \mathcal{P}(T)$ by:
$$\overline{\mu}(A) = \sup\left\{\sum_{i=1}^{n} |\mu(A_i)|;\ A_i \subset A,\ A_i \in \mathcal{C}, \forall i \in \{1, \ldots, n\}, A_i \cap A_j = \emptyset, \forall i \neq j\right\}$$
and
(ii) $|\mu| : \mathcal{C} \to \overline{\mathbb{R}}_+$, defined by $|\mu|(A) = |\mu(A)|$, for every $A \in \mathcal{C}$.

If μ is a multisubmeasure, then $\overline{\mu}$ is finitely additive and $|\mu|$ is a submeasure in the sense of Drewnowski [7].

3.2 Non-atomicity for set multifunctions

In this section, we extend the notion of a fuzzy multivalued set function (introduced by Guo and Zhang [13]) to the class of multivalued set functions taking values in $\mathcal{P}_f(X)$, the family of nonempty closed subsets of a real normed space X. We introduce the concepts of atom and non-atomicity for a multivalued set function and we point out some of their properties.

Definition 3.2.1 A multivalued set function $\mu : \mathcal{C} \to \mathcal{P}_f(X)$, with $\mu(\emptyset) = \{0\}$ is said to be *fuzzy* if μ satisfies the following conditions:

(i) μ is monotone;

(ii) μ is *increasing convergent* (i.e., $\mu(\bigcup_{n=1}^{\infty} A_n) = \lim_{n\to\infty} \mu(A_n)$ with respect to h, for every increasing sequence of sets $(A_n)_{n\in\mathbb{N}^*} \subset \mathcal{C}$, with $A = \bigcup_{n=1}^{\infty} A_n \in \mathcal{C}$, denoted by $A_n \nearrow A$);

(iii) μ is *decreasing convergent* (i.e., $\mu(\bigcap_{n=1}^{\infty} A_n) = \lim_{n\to\infty} \mu(A_n)$ with respect to h, for every decreasing sequence of sets $(A_n)_{n\in\mathbb{N}^*} \subset \mathcal{C}$, with $A = \bigcap_{n=1}^{\infty} A_n \in \mathcal{C}$, denoted by $A_n \searrow A$).

Remark 3.2.2

I. Guo and Zhang [13] introduced and studied the notion of fuzzy set-valued measure taking values in $\mathcal{P}_0(\overline{\mathbb{R}}_+)$. In their setting:

(a) A fuzzy set-valued measure is a function $\pi : \mathcal{A} \to \mathcal{P}_0(\overline{\mathbb{R}}_+)$ defined on a σ-algebra $\mathcal{A}$ of subsets of T and satisfying the conditions:

(i) $\pi(\emptyset) = \{0\}$.

(ii) $A \subset B$ implies $\pi(A) \leq \pi(B)$, for every $A, B \in \mathcal{A}$, i.e., π is monotone with respect to the relation "$\leq$", defined for every $E, F \in \mathcal{P}_0(\overline{\mathbb{R}}_+)$ by:

$$(*)\ E \leq F \text{ means } \begin{cases} \text{for each } x_0 \in E, \text{ there is } y_0 \in F, \text{ such that } x_0 \leq y_0 \text{ and} \\ \text{for each } y_0 \in F, \text{ there is } x_0 \in E, \text{ such that } x_0 \leq y_0. \end{cases}$$

(b) A fuzzy set-valued measure π is said to be continuous if π is:

(i) continuous from below, i.e., $A_n \nearrow A$ implies $\pi(A_n) \to \pi(A)$ and

(ii) continuous from above, i.e., $A_n \searrow A$ and there is $n_0 \in \mathbb{N}$, such that $\pi(A_{n_0}) \subset [0, +\infty)$ imply $\pi(A_n) \to \pi(A)$, where the convergence of $\pi(A_n)$ is in the Kuratowski sense (Klein and Thompson [16]). In

$\mathcal{P}_f(\overline{\mathbb{R}}_+)$, the Kuratowski convergence is equivalent to the convergence in Hausdorff topology. In general, the Kuratowski convergence is not topological and does not coincide with the Hausdorff one (see Section 1.5 in Chapter 1 for the definition and other considerations on the Kuratowski convergence).

Guo and Zhang [13] proved that the set-valued set function defined by:

$$\pi(A) = \int_A F d\mu = \{\int_A f d\mu;\ f \text{ is a measurable selection of } F\}, \forall A \in \mathcal{A},$$

is a fuzzy set-valued measure, where $\int_A f d\mu$ is Sugeno's fuzzy integral [25] of f with respect to a fuzzy measure μ and $F : T \to \mathcal{P}_0(\overline{\mathbb{R}}_+)$ is a multifunction.

II. According to Zhang, Guo and Liu [30], the set-valued set function defined by:

$$\pi(A) = (C)\int_A F d\mu = \{(C)\int_A f d\mu;\ f \in \mathcal{S}_C(F)\}, \forall A \in \mathcal{A},$$

is also a fuzzy set-valued measure, where $\mathcal{S}_C(F) = \{f;\ f$ is Choquet integrable and $f(t) \in F(t)$ μ-a.e.$\}$ and $(C)\int_A f d\mu$ is the Choquet integral of a measurable function f with respect to a fuzzy measure μ.

III. Kawabe [15] considered fuzzy measures as been continuous (from above and from below) monotone set functions $\mu : \mathcal{A} \to V$, where V is a Riesz space.

IV. In Definition 3.2.1, we considered the definitions of Guo and Zhang [13] and Kawabe [15] in the case of multivalued set functions μ having values in $\mathcal{P}_f(X)$, the family of nonempty closed subsets of a real normed space X. In this case, μ is monotone with respect to the ordinary inclusion of sets. In general, there is no implication between the relation $(*)$ of I and the inclusion on $\mathcal{P}_0(\overline{\mathbb{R}}_+)$ or $\mathcal{P}_f(\overline{\mathbb{R}}_+)$. Thus:

(i) Let be $E = [0, 2]$ and $F = [1, 4]$. Then $E \leq F$, but $E \not\subseteq F$.
(ii) Let be $E = [1, 2]$ and $F = [0, 4]$. Then $E \subseteq F$, but $E \not\leq F$.

Although, there are situations when they coincide, for example, on the family $\{[0, a]\ ;\ 0 \leq a < \infty\}$.

In our definition we chose this version because it seems to be more adequately for approaching and extending some topics of classical measure theory to the multivalued case (for certain applications in regularity, see Gavriluţ [11]).

We remark that Definition 3.2.1 cannot be reduced to that of Kawabe [15] because $\mathcal{P}_f(X)$ is not a linear space and, consequently, not a Riesz space (since $\mathcal{P}_f(X)$ is not a group with respect to the addition "+" defined by $M + N = \{x + y \mid x \in M, y \in N\}$, for every $M, N \in \mathcal{P}_f(X)$).

In Kawabe [15], in the case when the Riesz space V is the real line $\mathbb{R}$, his definition reduces to the usual fuzzy measure. In our setting, if $\mu : \mathcal{C} \to \mathcal{P}_f(\mathbb{R})$, with $\mu(\emptyset) = \{0\}$, is actually single-valued and monotone, then μ reduces in fact to zero function. So, Definition 3.2.1 does not reduce to the usual single-valued case and this new setting in the set-valued case could generate interesting questions and important problems.

Possibly Definition 3.2.1 might be a special case of Kawabe [15] if the set multifunctions took values in $\mathcal{P}_{bfc}(X)$. In this case, according to the embedding theorem of Rådström [23], $\mathcal{P}_{bfc}(X)$ might be embedded as a convex cone in a normed space V (but this delicate situation might complicate some settings).

Remark 3.2.3 Let $\mu : \mathcal{C} \to \mathcal{P}_f(X)$ be a multivalued set function, with $\mu(\emptyset) = \{0\}$. By the definitions, if μ is decreasing convergent, then μ is o-continuous.

Also, (Gavriluţ [8]) if μ is an o-continuous multisubmeasure, then μ is decreasing convergent and increasing convergent. Consequently, a multisubmeasure $\mu : \mathcal{C} \to \mathcal{P}_f(X)$ is o-continuous if and only if it is fuzzy.

Remark 3.2.4

(I) If $\mathcal{C}$ is finite, then every multivalued set function is increasing convergent and decreasing convergent.

(II) Let μ be the multisubmeasure induced by a submeasure ν, that is defined by $\mu(A) = [0, \nu(A)]$, for every $A \in \mathcal{C}$. Then μ is fuzzy if and only if ν is fuzzy. Indeed, for every $n \in \mathbb{N}$,

$$h(\mu(A_n), \mu(A)) = h([0, \nu(A_n)], [0, \nu(A)]) = |\nu(A_n) - \nu(A)|,$$

where $(A_n) \subset \mathcal{C}$, with $A_n \nearrow A \in \mathcal{C}$.

In Definition 3.2.1 we recalled the notions of an atom of a set function, a non-atomic set function, respectively.

We now introduce the concepts of an atom in the set-valued case and of a non-atomic multivalued set function.

Definition 3.2.5 Let $\mu : \mathcal{C} \to \mathcal{P}_0(X)$ be a multivalued set function, with $\mu(\emptyset) = \{0\}$.

(i) A set $A \in \mathcal{C}$ is said to be an *atom* of μ if $\mu(A) \supsetneq \{0\}$ and for every $B \in \mathcal{C}$, with $B \subset A$, we have $\mu(B) = \{0\}$ or $\mu(A \backslash B) = \{0\}$.

(ii) If μ is monotone, then μ is said to be *non-atomic* (NA) (or, *atomless*) if it has no atoms (that is, for every $A \in \mathcal{C}$, with $\mu(A) \supsetneq \{0\}$, there exists $B \in \mathcal{C}$, with $B \subset A$, $\mu(B) \supsetneq \{0\}$ and $\mu(A \backslash B) \supsetneq \{0\}$).

Example 3.2.6

(I) Let be $T = \{x, y, z\}$, $\mathcal{C} = \mathcal{P}(T)$ and for every $A \in \mathcal{C}$, let

$$\mu(A) = \begin{cases} [0, 1], & \text{if } A \neq \emptyset, \\ \{0\}, & \text{if } A = \emptyset. \end{cases}$$

One can easily check that $\mu : \mathcal{C} \to \mathcal{P}_f(\mathbb{R})$ is a multisubmeasure.

For $A = \{x, y\}$, there is $B = \{x\} \subset A$, such that $\mu(B) \supsetneq \{0\}$ and $\mu(A\backslash B) = \mu(\{y\}) = [0, 1] \supsetneq \{0\}$. So, A is not an atom of μ.

(II) Let be $\mathcal{C} = \{\emptyset, \{1\}, \{2\}, \{1, 2\}\}$ and the submeasure $\nu : \mathcal{C} \to \mathbb{R}_+$ defined by

$$\nu(A) = \begin{cases} 0, & \text{if } A = \emptyset \\ 1, & \text{if } A = \{1\} \text{ or } A = \{2\} \\ \frac{3}{2}, & \text{if } A = \{1, 2\}. \end{cases}$$

Then $\{1\}$ and $\{2\}$ are atoms for the multisubmeasure μ induced by ν.

(III) Suppose T is a countable set. Let be $\mathcal{C} = \{A;\ A \subset T,\ A$ is finite or cA is finite$\}$ and the multisubmeasure $\mu : \mathcal{C} \to \mathcal{P}_f(\mathbb{R})$, defined for every $A \in \mathcal{C}$ by:

$$\mu(A) = \begin{cases} \{0\}, & \text{if } A \text{ is finite} \\ \{0, 1\}, & \text{if } cA \text{ is finite} \end{cases}.$$

Then every $A \in \mathcal{C}$, such that cA is finite, is an atom of μ.

(IV) Let $\mathcal{C}$ be a ring of subsets of T, $m : \mathcal{C} \to \mathbb{R}_+$ a non-atomic, finitely additive set function and $\mu : \mathcal{C} \to \mathcal{P}_{bf}(\mathbb{R})$ the multivalued set function defined for every $A \in \mathcal{C}$ by

$$\mu(A) = \begin{cases} [-m(A), m(A)], & \text{if } m(A) \leq 1 \\ \\ [-m(A), 1], & \text{if } m(A) > 1 \end{cases}.$$

We easily observe that μ is a multisubmeasure. Since $|\mu| = m$, it follows that μ is non-atomic.

In what follows, we establish several properties of atoms and of non-atomic multivalued set functions.

Proposition 3.2.7

(I) Let $\nu : \mathcal{C} \to \mathbb{R}_+$ be a submeasure and μ, the multisubmeasure induced by ν. Then μ is non-atomic if and only if ν is non-atomic.

(II) If $\mu : \mathcal{C} \to \mathcal{P}_f(X)$ is monotone, with $\mu(\emptyset) = \{0\}$ and if $A \in \mathcal{C}$ is an atom of μ, then for every $B \in \mathcal{C}$, with $B \subset A$, we have either $\mu(B) = \{0\}$ or B is an atom of μ and $\mu(A\backslash B) = \{0\}$.

(III) If $\mu : \mathcal{C} \to \mathcal{P}_f(X)$ is a multisubmeasure and if $A \in \mathcal{C}$ is an atom of μ, then $\overline{\mu}(A) = |\mu(A)|$.

(IV) Let $\mu : \mathcal{C} \to \mathcal{P}_f(X)$ be monotone, with $\mu(\emptyset) = \{0\}$ and let $A \in \mathcal{C}$. The following statements are equivalent:

(i) A is an atom of μ;
(ii) A is an atom of $|\mu|$
(iii) A is an atom of $\overline{\mu}$.

(V) Let $\mu : \mathcal{C} \to \mathcal{P}_f(X)$ be a monotone multivalued set function, with $\mu(\emptyset) = \{0\}$. The following statements are equivalent:

(a) μ is non-atomic;
(b) $|\mu|$ is non-atomic
(c) $\overline{\mu}$ is non-atomic.

Proof (I), (II), (IV) and (V) are immediate consequences of the definitions.

(III) Let $A \in \mathcal{C}$ be an atom of μ. We always have $|\mu(A)| \leq \overline{\mu}(A)$. Now, let $\{B_i\}_{i=1}^n$ be a $\mathcal{C}$-partition of A. Then there is $i_0 \in \{1, \ldots, n\}$ such that $\mu(B_{i_0}) \supsetneq \{0\}$ and $\mu(B_i) = \{0\}$, for every $i \in \{1, \ldots, n\}, i \neq i_0$. Since μ is monotone, we have $\sum_{i=1}^{n} |\mu(B_i)| \leq |\mu(A)|$, which implies $\overline{\mu}(A) \leq |\mu(A)|$. □

Remark 3.2.8 If $\mu : \mathcal{C} \to \mathcal{P}_f(X)$ is an exhaustive multisubmeasure, then, according to Gavriluţ [8], μ takes its values in $\mathcal{P}_{bf}(X)$. Also, (Gavriluţ [8]) any o-continuous multisubmeasure defined on a σ-ring is exhaustive. Consequently, if $\mathcal{C}$ is a σ-ring and $\mu : \mathcal{C} \to \mathcal{P}_f(X)$ is a fuzzy multisubmeasure, then $\mu : \mathcal{C} \to \mathcal{P}_{bf}(X)$.

3.3 An extension by preserving non-atomicity

In this section we present an extension result that preserves non-atomicity, exhaustivity and the fuzzy character of a multisubmeasure/multimeasure.

Lemma 3.3.1 *Let $\mu : \mathcal{C} \to \mathcal{P}_f(X)$ be an exhaustive multivalued set function, with $\mu(\emptyset) = \{0\}$. Then for every $\varepsilon > 0$ and every $A \subset T$, there exists a set $K \in \mathcal{C}$ such that $K \subset A$ and $|\mu(B \backslash K)| < \varepsilon$, for every $B \in \mathcal{C}$, with $K \subset B \subset A$.*

Proof We suppose, on the contrary, that there are $\varepsilon_0 > 0$ and $A_0 \subset T$ so that for every $K \in \mathcal{C}$, with $K \subset A_0$, there exists $B_0 \in \mathcal{C}$ such that $K \subset B_0 \subset A_0$ and $|\mu(B_0 \backslash K)| \geq \varepsilon_0$.

We construct by induction a sequence of pairwise disjoint sets $(L_n)_n \subset \mathcal{C}$ so that $L_n \subset A_0$, for every $n \in \mathbb{N}$ and $|\mu(L_n)| \geq \varepsilon_0$. Since μ is exhaustive we shall have the contradiction.

Suppose that we have constructed $L_1, L_2, \ldots L_n$ and let $C = \bigcup_{i=1} n L_i$. Obviously, $C \in \mathcal{C}$ and $C \subset A_0$. Then there exists $B' \in \mathcal{C}$ so that $C \subset B' \subset A_0$ and $|\mu(B' \backslash C)| \geq \varepsilon_0$.

For $L_{n+1} = B' \backslash C$, we have $L_{n+1} \in \mathcal{C}$, $L_{n+1} \subset A_0$, $|\mu(L_{n+1})| \geq \varepsilon_0$ and $L_{n+1} \cap L_i = \emptyset$, for every $i \in \{1, \ldots, n\}$. □

From now on, suppose X is a Banach space.

Theorem 3.3.2 *Let $\mu : \mathcal{C} \to \mathcal{P}_f(X)$ be an exhaustive multisubmeasure. Then for every $A \subset T$, there exists in $\mathcal{P}_{bf}(X)$ the limit $\lim_{B \in \mathcal{C},\, B \subset A} \mu(B)$, denoted by $\mu^*(A)$, where $(\mu(B))_{B \in \mathcal{C},\, B \subset A}$ is a net with the set of indices directed by inclusion.*

Proof Since Lemma 3.3.1, for every $\varepsilon > 0$ and every $A \subset T$, there exists $B_0 = K \in \mathcal{C}$ such that $B_0 \subset A$ and for every $B \in \mathcal{C}$, with $B \subset A$ and $B \supseteq B_0$, we have

$$h(\mu(B), \mu(B_0)) \leq |\mu(B \backslash B_0)| < \varepsilon.$$

Hence $(\mu(B))_{B \in \mathcal{C},\, B \subset A}$ is a Cauchy net in $\mathcal{P}_{bf}(X)$. Since $\mathcal{P}_{bf}(X)$ is complete with respect to h, the net $(\mu(B))_{B \in \mathcal{C},\, B \subset A}$ is convergent in $\mathcal{P}_{bf}(X)$. Its limit exists in $\mathcal{P}_{bf}(X)$ and is unique. □

In the sequel, we prove that any fuzzy multisubmeasure/multimeasure on a σ-algebra can be uniquely extended (also by preserving non-atomicity) to a wider family of sets. In order to do this, we need some other results:

Theorem 3.3.3 *Let $\mu : \mathcal{C} \to \mathcal{P}_f(X)$ be an exhaustive multisubmeasure. Then:*

(i) $\mu^*(A) = \mu(A)$, *for every* $A \in \mathcal{C}$.
(ii) μ^* *is a monotone multivalued set function, with* $\mu^*(\emptyset) = \{0\}$.
(iii) If μ *is non-atomic, then* μ^* *is also non-atomic.*
(iv) μ^* *is exhaustive.*

Proof

(i) For every $\varepsilon > 0$ and every $A \subset T$, there exists $B_0 \in \mathcal{C}$ such that $B_0 \subset A$ and for every $B \in \mathcal{C}$, with $B_0 \subset B \subset A$, we have $h(\mu(B), \mu^*(A)) < \varepsilon$. For $B = A \in \mathcal{C}$, we get that $h(\mu(A), \mu^*(A)) < \varepsilon$, for every $\varepsilon > 0$, hence $\mu^*(A) = \mu(A)$, for every $A \in \mathcal{C}$.

(ii) Obviously, by (i), $\mu^*(\emptyset) = \mu(\emptyset) = \{0\}$. Let $A_1, A_2 \subset T$, with $A_1 \subset A_2$. We prove that $\mu^*(A_1) \subseteq \mu^*(A_2)$. Let $\varepsilon > 0$ be arbitrary.

For A_1 there exists $B_1(\varepsilon) \in \mathcal{C}$ so that $B_1 \subset A_1$ and for every $B \in \mathcal{C}$, with $B_1 \subset B \subset A_1$, we have $h(\mu^*(A_1), \mu(B)) < \frac{\varepsilon}{2}$.

Analogously, for A_2 there exists $B_2(\varepsilon) \in \mathcal{C}$ so that $B_2 \subset A_2$ and for every $B \in \mathcal{C}$, with $B_2 \subset B \subset A_2$, we have $h(\mu^*(A_2), \mu(B)) < \frac{\varepsilon}{2}$.

Let be $B_0 = B_1 \cup B_2$. Then $B_2 \subset B_0 \subset A_2$, hence $h(\mu^*(A_2), \mu(B_0)) < \frac{\varepsilon}{2}$. Consequently,

$$e(\mu^*(A_1), \mu^*(A_2)) < \varepsilon + e(\mu(B_1), \mu(B_0)) = \varepsilon.$$

Because ε is arbitrary, the conclusion follows.

(iii) Suppose that, on the contrary, there exists an atom $A_0 \in \mathcal{P}(T)$ for μ^*. Then $\mu^*(A_0) \supsetneq \{0\}$ and for every $B \subset T$, with $B \subset A_0$, we have that $\mu^*(B) = \{0\}$ or $\mu^*(A_0 \backslash B) = \{0\}$. Because $\mu^*(A_0) \supsetneq \{0\}$, there exists $C_0 \in \mathcal{C}$ so that $C_0 \subset A_0$ and $\mu(C_0) \supsetneq \{0\}$.

Since μ is non-atomic, there is $D_0 \in \mathcal{C}$ so that $D_0 \subset C_0$, $\mu(D_0) \supsetneq \{0\}$ and $\mu(C_0 \backslash D_0) \supsetneq \{0\}$. For D_0, $\mu^*(D_0) = \{0\}$ or $\mu^*(A_0 \backslash D_0) = \{0\}$.

If $\mu^*(D_0) = \{0\}$, then

$$\mu(D_0) = \mu^*(D_0) = \{0\},$$

which is false.

If $\mu^*(A_0 \backslash D_0) = \{0\}$, then

$$\{0\} \subsetneq \mu(C_0 \backslash D_0) = \mu^*(C_0 \backslash D_0) \subseteq \mu^*(A_0 \backslash D_0) = \{0\},$$

a contradiction. So, μ^* is non-atomic, as claimed.

(iv) Let be $(A_n)_n \subset T$, $A_n \cap A_m = \emptyset$, $m \neq n$.

For every $\varepsilon > 0$ and every $n \in \mathbb{N}$, there exists $B_n(\varepsilon) \in \mathcal{C}$ so that $B_n \subset A_n$ and $h(\mu^*(A_n), \mu(B_n)) < \frac{\varepsilon}{2}$.

Since $B_n \cap B_m = \emptyset$, $m \neq n$ and μ is exhaustive, then there exists $n_0(\varepsilon) \in \mathbb{N}$ such that $|\mu(B_n)| < \frac{\varepsilon}{2}$, for every $n \geq n_0$. Therefore, for every $\varepsilon > 0$, there exists $n_0(\varepsilon) \in \mathbb{N}$ such that for every $n \geq n_0$,

$$|\mu^*(A_n)| \leq h(\mu^*(A_n), \mu(B_n)) + |\mu(B_n)| < \varepsilon.$$

Consequently, $\lim_{n \to \infty} |\mu^*(A_n)| = 0$, which means that μ^* is exhaustive. □

From now on, $\mathcal{C}$ is an algebra of subsets of T.

Lemma 3.3.4 *Let $\mu : \mathcal{C} \to \mathcal{P}_f(X)$ be an exhaustive multisubmeasure. Then, for every $\varepsilon > 0$ and every $A \subset T$, there exists a set $D \in \mathcal{C}$ such that $A \subset D$ and $|\mu(D \backslash B)| < \varepsilon$, for every $B \in \mathcal{C}$, with $A \subset B \subset D$.*

Proof We use Lemma 3.3.1 for cA : for every $\varepsilon > 0$, there exists $K \in \mathcal{C}$ such that $K \subset cA$ and $|\mu(B \backslash K)| < \varepsilon$, for every $B \in \mathcal{C}$, with $K \subset B \subset cA$. Let $D = cK$. Then $D \in \mathcal{C}$, $A \subset D$ and for every $B \in \mathcal{C}$, with $A \subset B \subset D$, we get that $K \subset cB \subset cA$, hence $|\mu(D \backslash B)| = |\mu(cB \backslash K)| < \varepsilon$. □

Now, let be $\mathcal{C}_\mu = \{A \subset T$; for every $\varepsilon > 0$, there exist $K, D \in \mathcal{C}$ such that $K \subset A \subset D$ and $|\mu(B)| < \varepsilon$, for every $B \in \mathcal{C}$, with $B \subset D \backslash K\}$.

Obviously, $\mathcal{C} \subset \mathcal{C}_\mu$ and it is easy to check that $\mathcal{C}_\mu$ is an algebra.

Theorem 3.3.5 *Let $\mu : \mathcal{C} \to \mathcal{P}_f(X)$ be an exhaustive multisubmeasure. If $A \in \mathcal{C}_\mu$, then:*

(i) For every $\varepsilon > 0$, there exist $K, D \in \mathcal{C}$ such that $K \subset A \subset D$ and $h(\mu(B'), \mu(B'')) < \frac{\varepsilon}{2}$, for every $B', B'' \in \mathcal{C}$, with $K \subset B', B'' \subset D$.

(ii) $h(\mu^*(A), \mu(B)) < \varepsilon$, for every $B \in \mathcal{C}$, with $K \subset B \subset D$.

Proof

(i) Let be $B', B'' \in \mathcal{C}$, with $K \subset B', B'' \subset D$. Then $B' \backslash B'', B'' \backslash B' \subset D \backslash K$, hence $B' \Delta B'' \subset D \backslash K$. Consequently,

$$h(\mu(B'), \mu(B'')) \leq |\mu(B' \Delta B'')| < \frac{\varepsilon}{2}.$$

(ii) Let be $B \in \mathcal{C}$, with $K \subset B \subset D$, be arbitrary, but fixed. By (i), for every $B' \in \mathcal{C}$, with $K \subset B' \subset A$, we have $h(\mu(B'), \mu(B)) < \frac{\varepsilon}{2}$.

By the definition of μ^*, there is $B_0 \in \mathcal{C}$ such that $B_0 \subset A$ and for every $\widetilde{B} \in \mathcal{C}$, with $B_0 \subset \widetilde{B} \subset A$, we have $h(\mu(\widetilde{B}), \mu^*(A)) < \frac{\varepsilon}{2}$.

Let be $\widetilde{B} \cup B'$. Then $K \subset \widetilde{B} \cup B' \subset A$ and $B_0 \subset \widetilde{B} \cup B' \subset A$, which imply that

$$h(\mu(\widetilde{B} \cup B'), \mu(B)) < \frac{\varepsilon}{2} \text{ and } h(\mu(\widetilde{B} \cup B'), \mu^*(A)) < \frac{\varepsilon}{2}.$$

Consequently,

$$h(\mu^*(A), \mu(B)) \leq h(\mu(\widetilde{B} \cup B'), \mu(B)) + h(\mu(\widetilde{B} \cup B'), \mu^*(A)) < \varepsilon,$$

for every $B \in \mathcal{C}$, with $K \subset B \subset D$, as claimed. □

Theorem 3.3.6 *Let $\mu : \mathcal{C} \to \mathcal{P}_f(X)$ be an exhaustive multisubmeasure. Then:*

(i) $\mu^*_{/\mathcal{C}_\mu}$ is an exhaustive multisubmeasure which extends μ. Moreover, if μ is non-atomic (a multimeasure respectively), then the same is $\mu^*_{/\mathcal{C}_\mu}$.

(ii) $\mu^*(A_1 \cup A_2) \subseteq \mu^*(A_1) \overset{\bullet}{+} \mu^*(A_2)$, for every $A_1, A_2 \in \mathcal{P}(T)$ such that there is $C \in \mathcal{C}_\mu$, $A_1 \subset C$ and $A_2 \cap C = \emptyset$.

(iii) $\mu^*(A) \subseteq \mu^*(A \backslash B) \overset{\bullet}{+} \mu^*(B)$, for every $A, B \in \mathcal{P}(T)$, with $B \in \mathcal{C}$, $B \subset A$.

Proof

(i) Since μ^* is a monotone, exhaustive multivalued set function so that $\mu^*(\emptyset) = \{0\}$, then the same is $\mu^*_{/\mathcal{C}_\mu}$. Also, the same as in the proof of Theorem 3.3.3(iii), we have that $\mu^*_{/\mathcal{C}_\mu}$ is non-atomic.

Now, let be $A_1, A_2 \in \mathcal{C}_\mu$, with $A_1 \cap A_2 = \emptyset$. We prove that $\mu^*(A_1 \cup A_2) \subseteq \mu^*(A_1) \overset{\bullet}{+} \mu^*(A_2)$. Indeed, since $A_1 \cup A_2 \in \mathcal{C}_\mu$, then there are $K, D \in \mathcal{C}$ such that $K \subset A_1 \cup A_2 \subset D$ and $h(\mu^*(A_1 \cup A_2), \mu(B)) < \frac{\varepsilon}{3}$, for every $B \in \mathcal{C}$, with $K \subset B \subset D$.

Analogously, for $A_1 \in \mathcal{C}_\mu$ there are $K_1, D_1 \in \mathcal{C}$ such that $K_1 \subset A_1 \subset D_1$ and $h(\mu^*(A_1), \mu(B)) < \frac{\varepsilon}{3}$, for every $B \in \mathcal{C}$, with $K_1 \subset B \subset D_1$ and for $A_2 \in \mathcal{C}_\mu$ there are $K_2, D_2 \in \mathcal{C}$ such that $K_2 \subset A_2 \subset D_2$ and $h(\mu^*(A_2), \mu(B)) < \frac{\varepsilon}{3}$, for every $B \in \mathcal{C}$, with $K_2 \subset B \subset D_2$.

Let be $\widetilde{K}_1 = K_1 \cup [(K \backslash K_2) \cap D_1] \in \mathcal{C}$ and $\widetilde{K}_2 = (D_2 \cap D) \backslash \widetilde{K} \in \mathcal{C}$. One can easily check that $\widetilde{K}_1 \cap \widetilde{K}_2 = \emptyset$, $K_1 \subset \widetilde{K}_1 \subset D_1$, $K_2 \subset \widetilde{K}_2 \subset D_2$ and $K \subset \widetilde{K}_1 \cup \widetilde{K}_2 \subset D$. Therefore,

$$h(\mu^*(A_1), \mu(\widetilde{K}_1)) < \frac{\varepsilon}{3}, h(\mu^*(A_2), \mu(\widetilde{K}_2)) < \frac{\varepsilon}{3} \text{ and}$$

$$h(\mu^*(A_1 \cup A_2), \mu(\widetilde{K}_1 \cup \widetilde{K}_2)) < \frac{\varepsilon}{3}.$$

Because $\mu(\widetilde{K}_1 \cup \widetilde{K}_2) \subseteq \mu(\widetilde{K}_1) \overset{\bullet}{+} \mu(\widetilde{K}_2)$, then $e(\mu(\widetilde{K}_1 \cup \widetilde{K}_2), \mu(\widetilde{K}_1) \overset{\bullet}{+} \mu(\widetilde{K}_2)) = 0$. We immediately get that $e(\mu^*(A_1 \cup A_2), \mu^*(A_1) \overset{\bullet}{+} \mu^*(A_2)) < \varepsilon$, for every $\varepsilon > 0$, hence $\mu^*(A_1 \cup A_2) \subseteq \mu^*(A_1) \overset{\bullet}{+} \mu^*(A_2)$, as claimed.

Now, we observe that, if, particularly, μ is a multimeasure, then instead of $\mu(\widetilde{K}_1 \cup \widetilde{K}_2) \subseteq \mu(\widetilde{K}_1) \overset{\bullet}{+} \mu(\widetilde{K}_2)$, we have $\mu(\widetilde{K}_1 \cup \widetilde{K}_2) = \mu(\widetilde{K}_1) \overset{\bullet}{+} \mu(\widetilde{K}_2)$. The same as before, we get that

$$\begin{aligned} h(\mu^*(A_1 \cup A_2), \mu^*(A_1) \overset{\bullet}{+} \mu^*(A_2)) &\leq h(\mu^*(A_1 \cup A_2), \mu(\widetilde{K}_1 \cup \widetilde{K}_2)) \\ &+ h(\mu(\widetilde{K}_1 \cup \widetilde{K}_2), \mu(\widetilde{K}_1) \overset{\bullet}{+} \mu(\widetilde{K}_2)) \\ &+ h(\mu(\widetilde{K}_1) \overset{\bullet}{+} \mu(\widetilde{K}_2), \mu^*(A_1) \overset{\bullet}{+} \mu^*(A_2)) \\ &= h(\mu^*(A_1 \cup A_2), \mu(\widetilde{K}_1 \cup \widetilde{K}_2)) \\ &+ h(\mu(\widetilde{K}_1) \overset{\bullet}{+} \mu(\widetilde{K}_2), \mu^*(A_1) \overset{\bullet}{+} \mu^*(A_2)) < \varepsilon, \end{aligned}$$

for every $\varepsilon > 0$, hence $\mu^*(A_1 \cup A_2) = \mu^*(A_1) \overset{\bullet}{+} \mu^*(A_2)$, as claimed.

(ii) By the definition of μ^*, there are $K, K_1, K_2 \in \mathcal{C}$ so that $K_1 \subset A_1, K_2 \subset A_2, K \subset A_1 \cup A_2$ and

$h(\mu^*(A_1 \cup A_2), \mu(M)) < \frac{\varepsilon}{3}$, for every $M \in \mathcal{C}$, with $K \subset M \subset A_1 \cup A_2$,
$h(\mu^*(A_1), \mu(M)) < \frac{\varepsilon}{3}$, for every $M \in \mathcal{C}$, with $K_1 \subset M \subset A_1$ and
$h(\mu^*(A_2), \mu(M)) < \frac{\varepsilon}{3}$, for every $M \in \mathcal{C}$, with $K_2 \subset M \subset A_2$.

By the hypothesis, there exists $C \in \mathcal{C}$ such that $A_1 \subset C$ and $A_2 \cap C = \emptyset$. Let $D_1 = K_1 \cup (K \cap C)$ and $D_2 = K_2 \cup (K \backslash C)$. Obviously, $D_1, D_2 \in \mathcal{C}$, $K_1 \subset D_1 \subset A_1$, $K_2 \subset D_2 \subset A_2$, $K \subset D_1 \cup D_2 \subset A_1 \cup A_2$ and $D_1 \cap D_2 = \emptyset$. Then

$$h(\mu^*(A_1 \cup A_2), \mu(D_1 \cup D_2)) < \frac{\varepsilon}{3}, h(\mu^*(A_1), \mu(D_1)) < \frac{\varepsilon}{3} \text{ and}$$

$$h(\mu^*(A_2), \mu(D_2)) < \frac{\varepsilon}{3}.$$

Because $\mu(D_1 \cup D_2) \subseteq \mu(D_1) \overset{\bullet}{+} \mu(D_2)$, we get that

$$e((\mu^*(A_1 \cup A_2), \mu^*(A_1) \overset{\bullet}{+} \mu^*(A_2)) < \varepsilon, \text{ for every } \varepsilon > 0,$$

hence $\mu^*(A_1 \cup A_2) \subseteq \mu^*(A_1) \overset{\bullet}{+} \mu^*(A_2)$.

(iii) In (ii) put $A_1 = B$, $A_2 = A \backslash B$ and $C = B$. □

We are now able to establish our main extension result.

Theorem 3.3.7 *Let $\mathcal{C}$ be a σ-algebra and $\mu : \mathcal{C} \to \mathcal{P}_{bf}(X)$ a fuzzy multisubmeasure (monotone multimeasure, respectively). Then μ uniquely extends to a fuzzy multisubmeasure (monotone multimeasure, respectively) $\mu^* : \mathcal{C}_\mu \to \mathcal{P}_{bf}(X)$.*

Proof Since μ is fuzzy, then, according to Remarks 3.2.3 and 3.2.8, $\mu : \mathcal{C} \to \mathcal{P}_{bf}(X)$ and μ is exhaustive. By Theorem 3.3.6, μ extends to a multisubmeasure (monotone multimeasure, respectively) $\mu^* : \mathcal{C}_\mu \to \mathcal{P}_{bf}(X)$.

We prove that μ^* is fuzzy, or, equivalently, by Remark 3.2.3, that μ^* is o-continuous. Let then $(A_n)_{n\in\mathbb{N}^*} \subset \mathcal{C}_\mu$, with $A_n \searrow \emptyset$. For $n = 1$, there are $K_1, E_1 \in \mathcal{C}$ such that $K_1 \subset A_1 \subset E_1$ and $|\mu(B)| < \frac{\varepsilon}{2}$, for every $B \in \mathcal{C}$, with $B \subset E_1\backslash K_1$. Denote $D_1 = E_1$. Then, particularly, $|\mu(D_1\backslash K_1)| < \frac{\varepsilon}{2}$.

For $n = 2$, there are $K_2, E_2 \in \mathcal{C}$ such that $K_2 \subset A_2 \subset E_2$ and $|\mu(B)| < \frac{\varepsilon}{2^2}$, for every $B \in \mathcal{C}$, with $B \subset E_2\backslash K_2$. Denote $D_2 = D_1 \cap E_2$. We have $D_2 \in \mathcal{C}$, $D_2 \subset D_1$ and, particularly, $|\mu(D_2\backslash K_2)| < \frac{\varepsilon}{2^2}$. Recurrently, there are $(K_n)_{n\in\mathbb{N}^*} \subset \mathcal{C}$, $(D_n)_{n\in\mathbb{N}^*} \subset \mathcal{C}$, such that $K_n \subset A_n \subset D_n$, $D_n \supset D_{n+1}$ and $|\mu(D_n\backslash K_n)| < \frac{\varepsilon}{2^n}$, for every $n \in \mathbb{N}^*$.

Let be $\widetilde{K}_n = \bigcap_{i=1}^{n} K_i$, for every $n \in \mathbb{N}^*$. Then $\widetilde{K}_n \searrow \emptyset$ and, since μ is fuzzy (equivalently, o-continuous) on $\mathcal{C}$, we get that $\lim_{n\to\infty} |\mu(\widetilde{K}_n)| = 0$.

Consequently, because $D_n\backslash\widetilde{K}_n \subseteq \bigcup_{n}^{i=1}(D_i\backslash K_i)$, we have:

$$\begin{aligned}
|\mu^*(A_n)| &\le |\mu^*(A_n\backslash\widetilde{K}_n)| + |\mu(\widetilde{K}_n)| \le |\mu^*(D_n\backslash\widetilde{K}_n)| + |\mu(\widetilde{K}_n)| \\
&= |\mu(D_n\backslash\widetilde{K}_n)| + |\mu(\widetilde{K}_n)| \le \sum_{i=1}^{n} |\mu(D_i\backslash K_i)| + |\mu(\widetilde{K}_n)| \\
&< \sum_{i=1}^{n} \frac{\varepsilon}{2^i} + |\mu(\widetilde{K}_n)| < \varepsilon + |\mu(\widetilde{K}_n)|, \text{ for every } n \in \mathbb{N}^*.
\end{aligned}$$

Since $\lim_{n\to\infty} |\mu(\widetilde{K}_n)| = 0$, we get that $\lim_{n\to\infty} |\mu^*(A_n)| = 0$, hence μ^* is o-continuous.

Let us prove now that the extension μ^* is unique. Suppose that, on the contrary, there exists another fuzzy multisubmeasure (monotone multimeasure, respectively) $\nu : \mathcal{C}_\mu \to \mathcal{P}_{bf}(X)$ which extends μ. Let $A \in \mathcal{C}_\mu$ be arbitrary. By the definition of $\mathcal{C}_\mu$, for every $\varepsilon > 0$, there are $K, D \in \mathcal{C}$ so that $K \subset A \subset D$ and $|\mu(D\backslash K)| < \varepsilon$.

Then, for every $\varepsilon > 0$,

$$\begin{aligned}
e(\mu^*(A), \nu(A)) &\le e(\mu^*(A), \mu^*(D)) + e(\mu^*(D), \nu(A)) = e(\mu(D), \nu(A)) \\
&\le e(\mu(D), \mu(K)) + e(\mu(K), \nu(A)) = e(\mu(D), \mu(K)) \\
&\le |\mu(D\backslash K)| < \varepsilon,
\end{aligned}$$

hence $\mu^*(A) \subseteq \nu(A)$. Analogously, $\nu(A) \subseteq \mu^*(A)$ and the conclusion follows. □

Remark 3.3.8 In general, if $\mathcal{C}_1 \subset \mathcal{C}_2$ are rings and if $\mu : \mathcal{C}_2 \to \mathcal{P}_{bf}(X)$ is a non-atomic multisubmeasure on $\mathcal{C}_2$, we do not always have that μ is non-atomic on $\mathcal{C}_1$. Although, if $\mathcal{C}_1$ is dense in $\mathcal{C}_2$ with respect to a multisubmeasure μ (that is, for every

$\varepsilon > 0$ and every $A \in \mathcal{C}_2$, there exists $B \in \mathcal{C}_1$ so that $B \subset A$ and $|\mu(A \backslash B)| < \varepsilon$) and if μ is non-atomic on $\mathcal{C}_2$, then μ is also non-atomic on $\mathcal{C}_1$.

Indeed, suppose that, on the contrary, there exists an atom $A \in \mathcal{C}_1$ for $\mu_{/\mathcal{C}_1}$. Then $\mu(A) \supsetneq \{0\}$ and for every $B \in \mathcal{C}_1$ with $B \subset A$ we have that $\mu(B) = \{0\}$ or $\mu(A \backslash B) = \{0\}$.

Because $A \in \mathcal{C}_2$, $\mu(A) \supsetneq \{0\}$ and μ is non-atomic on $\mathcal{C}_2$, there is $B_0 \in \mathcal{C}_2$ so that $B_0 \subset A$, $\mu(B_0) \supsetneq \{0\}$ and $\mu(A \backslash B_0) \supsetneq \{0\}$. Then $|\mu(B_0)| > 0$ and, since $\mathcal{C}_1$ is dense in $\mathcal{C}_2$, then for $\varepsilon = |\mu(B_0)|$, there exists $C_0 \in \mathcal{C}_1$ so that $C_0 \subset B_0$ and $|\mu(B_0 \backslash C_0)| < \varepsilon_0$.

Now, because $C_0 \in \mathcal{C}_1$ and $C_0 \subset A$, by the assumption we made we get that $\mu(C_0) = \{0\}$ or $\mu(A \backslash C_0) = \{0\}$.

(I) If $\mu(C_0) = \{0\}$, then

$$|\mu(B_0)| \leq |\mu(B_0 \backslash C_0)| + |\mu(C_0)| = |\mu(B_0 \backslash C_0)| < |\mu(B_0)|,$$

which is false.

(II) If $\mu(A \backslash C_0) = \{0\}$, then

$$|\mu(A \backslash B_0)| \leq |\mu(A \backslash C_0)| = 0,$$

hence $\mu(A \backslash B_0) = \{0\}$, which is again false.

So, μ is non-atomic on $\mathcal{C}_1$.

Remark 3.3.9 $\mathcal{C}$ is dense in $\mathcal{C}_\mu$ with respect to μ^*. Indeed, for every $\varepsilon > 0$ and every $A \in \mathcal{C}_\mu$, there exists $B, D \in \mathcal{C}$ so that $B \subset A \subset D$ and $|\mu(S)| < \varepsilon$, for every $S \in \mathcal{C}$ with $S \subset D \backslash B$. Particularly, $|\mu(D \backslash B)| < \varepsilon$. Then

$$|\mu^*(A \backslash B)| \leq |\mu^*(D \backslash B)| = |\mu(D \backslash B)| < \varepsilon.$$

Corollary 3.3.10

(i) μ is non-atomic on $\mathcal{C}$ if and only if μ^* is non-atomic on $\mathcal{C}_\mu$;
(ii) μ is fuzzy on $\mathcal{C}$ if and only if μ^* is fuzzy on $\mathcal{C}_\mu$.

Proof

(i) The "*Only if part*" follows from Theorem 3.3.6(i) and the "*If part*" is a consequence of Remarks 3.3.8 and 3.3.9.
(ii) This is an immediate consequence of Theorem 3.3.7. □

References

1. Aumann, R.J., Shapley, L.S.: Values of Non-atomic Games. Princeton University Press, Princeton (1974)
2. Choquet, G.: Theory of capacities. Ann. Inst. Fourier (Grenoble) **5**, 131–292 (1953–1954)

3. Dempster, A.P.: Upper and lower probabilities induced by a multivalued mapping. Ann. Math. Stat. **38**, 325–339 (1967)
4. Denneberg, D.: Non-additive Measure and Integral. Kluwer Academic Publishers, Dordrecht (1994)
5. Dinculeanu, N.: Teoria Măsurii şi Funcţii Reale (in Romanian). Didactic and Pedagogical Publishing House, Bucharest (1964)
6. Dobrakov, I.: On submeasures, I. Diss. Math. **112**, 5–35 (1974)
7. Drewnowski, L.: Topological rings of sets, continuous set functions. Integration, I, II, III. Bull. Acad. Pol. Sci. Sér. Math. Astron. Phys. **20**, 269–286 (1972)
8. Gavriluţ, A.: Properties of regularity for multisubmeasures. Ann. Şt. Univ. Iaşi Tomul L s. I a f. **2**, 373–392 (2004)
9. Gavriluţ, A.: Regularity and o-continuity for multisubmeasures. Ann. Şt. Univ. Iaşi Tomul L s. I a f. **2**, 393–406 (2004)
10. Gavriluţ, A.: $\mathcal{K}$-tight multisubmeasures, $\mathcal{K}-\mathcal{D}$-regular multisubmeasures. Ann. Şt. Univ. Iaşi Tomul LI s. I f. **2**, 387–404 (2005)
11. Gavriluţ, A.: Non-atomicity and the Darboux property for fuzzy and non-fuzzy Borel/Baire multivalued set functions. Fuzzy Sets Syst. **160**, 1308–1317 (2009)
12. Gavriluţ, A., Croitoru, A.: Pseudo-atoms and Darboux property for fuzzy and non-fuzzy set multifunctions. Fuzzy Sets Syst. **161**, 2897–2908 (2010)
13. Guo, C., Zhang, D.: On set-valued fuzzy measures. Inf. Sci. **160**, 13–25 (2004)
14. Hu, S., Papageorgiou, N.S.: Handbook of Multivalued Analysis, vol. I. Kluwer Academic Publishers, Dordrecht (1997)
15. Kawabe, J.: Regularity and Lusin's theorem for Riesz space-valued fuzzy measures. Fuzzy Sets Syst. **158**, 895–903 (2007)
16. Klein, E., Thompson, A.: Theory of Correspondences. Wiley, New York (1984)
17. Liginlal, O., Ow, T.T.: Modeling attitude to risk in human decision processes: an application of fuzzy measures. Fuzzy Sets Syst. **157**, 3040–3054 (2006)
18. Lipecki, Z.: Extensions of additive set functions with values in a topological group. Bull. Acad. Pol. Sci. **22**(1), 19–27 (1974)
19. Pap, E.: Regular null-additive monotone set function. Novi Sad J. Math. **25**(2), 93–101 (1995)
20. Pap, E.: Null-Additive Set Functions. Kluwer Academic Publishers, Dordrecht (1995)
21. Pham, T.D., Brandl, M., Nguyen, N.D., Nguyen, T.V.: Fuzzy measure of multiple risk factors in the prediction of osteoporotic fractures. In: Proceedings of the 9th WSEAS International Conference on Fuzzy Systems [FS'08], Sofia, Bulgaria, May 2–4, pp. 171–177 (2008)
22. Precupanu, A.M.: On the set valued additive and subadditive set functions. Ann. Şt. Univ. Iaşi **29**, 41–48 (1984)
23. Rådström, H.: An embedding theorem for spaces of convex sets. Proc. Amer. Math. Soc. **3**, 165–169 (1952)
24. Shafer, G.: A Mathematical Theory of Evidence. Princeton University Press, Princeton (1976)
25. Sugeno, M.: Theory of fuzzy integrals and its applications. Ph.D. Thesis, Tokyo Institute of Technology (1974)
26. Suzuki, H.: Atoms of fuzzy measures and fuzzy integrals. Fuzzy Sets Syst. **41**, 329–342 (1991)
27. Wu, C., Bo, S.: Pseudo-atoms of fuzzy and non-fuzzy measures. Fuzzy Sets Syst. **158**, 1258–1272 (2007)
28. Zhang, D., Wang, Z.: On set-valued fuzzy integrals. Fuzzy Sets Syst. **56**, 237–241 (1993)
29. Zhang, D., Guo, C.: Generalized fuzzy integrals of set-valued functions. Fuzzy Sets Syst. **76**, 365–373 (1995)
30. Zhang, D., Guo, C., Liu D.: Set-valued Choquet integrals revisited. Fuzzy Sets Syst. **147**, 475–485 (2004)

Chapter 4
Non-atomicity and the Darboux property for fuzzy and non-fuzzy Borel/Baire multivalued set functions

4.1 Introduction

In this chapter, we establish results concerning non-atomicity and the Darboux property for multisubmeasures/fuzzy multivalued set functions defined on the Baire (respectively, Borel) δ-ring $\mathcal{B}_0$ (respectively, $\mathcal{B}$) of a locally compact, Hausdorff space and taking values in the family $\mathcal{P}_f(X)$ of all nonvoid closed subsets of a real normed space X, endowed with the Hausdorff topology.

Regularity has been intensively studied for measures by Halmos [11], Dinculeanu [2], Kluvánek [3], Lewis [4], Khurana [12, 13], Belley [1], Sundaresan [21], Morales [15] and many others. Different problems concerning regularity have been solved by Precupanu [19] for multimeasures. One of the major interests was to prove the equivalence under some conditions between regularity and countable additivity. In [6, 7] and [8] we have introduced and studied different types of regularity for a special type of multivalued set functions (called in [6] a "multisubmeasure").

On the other hand, because of their multiple applications, for instance in non-atomic game theory, the study of non-atomic additive, respectively, non-additive set functions has been developed. In this sense, we mention the contributions of Dinculeanu [2], Dobrakov [5], Klimkin and Svistula [14], Olejcek [16], Pap [17, 18], Suzuki [22], Wu and Bo [23] and others. That is why, in [10], we defined non-atomicity and the Darboux property in the multivalued case. We have pointed out the relationships between non-atomicity and the Darboux property and we have also indicated the differences which appear here from the case of additive, respectively, non-additive set functions. Precisely, for σ-finite measures on a δ-ring, and for Dobrakov submeasures on a σ-algebra, non-atomicity and the Darboux property are equivalent ([14]). In the multivalued case, there appear differences: for multisubmeasures we have in general only the implication "Darboux property $\rightarrow$ non-atomicity". Although, as we shall see, for multisubmeasures induced by Dobrakov submeasures on a σ-algebra (see [10]), the equivalence still holds.

A. Gavriluţ et al., *Atomicity through Fractal Measure Theory*,
https://doi.org/10.1007/978-3-030-29593-6_4

4.2 Preliminary definitions and remarks

We firstly list the notations and several definitions used throughout this chapter.

Let T be a locally compact Hausdorff space, $\mathcal{C}$ a ring of subsets of T, $\mathcal{B}_0$ the Baire δ-ring, $\mathcal{B}$ the Borel δ-ring (see Section 2.3), X a real normed space (with the origin 0).

On $\mathcal{P}_0(X)$ we consider the Minkowski addition "$\dot{+}$" from Chapter 3 (see also [20]).

We recall from [6] the following notions (I)–(VI) and introduce the notion of a fuzzy multivalued set function:

Definition 4.2.1 Let $\mu : \mathcal{C} \to \mathcal{P}_f(X)$ be a multivalued set function, with $\mu(\emptyset) = \{0\}$.

(I) μ is said to be *order-continuous* (briefly, *o-continuous)* with respect to h if $\lim_{n\to\infty} |\mu(A_n)| = 0$, for every sequence of sets $(A_n)_{n\in\mathbb{N}^*} \subset \mathcal{C}$, with $A_n \searrow \emptyset$;

(II) μ is said to be *exhaustive* with respect to h if $\lim_{n\to\infty} |\mu(A_n)| = 0$, for every pairwise disjoint sequence of sets $(A_n)_{n\in\mathbb{N}^*} \subset \mathcal{C}$;

(III) μ is said to be *decreasing convergent* with respect to h if $\lim_{n\to\infty} h(\mu(A_n), \mu(A)) = 0$, for every sequence of sets $(A_n)_{n\in\mathbb{N}^*} \subset \mathcal{C}$, with $A_n \searrow A \in \mathcal{C}$;

(III') μ is said to be *increasing convergent* with respect to h if $\lim_{n\to\infty} h(\mu(A_n), \mu(A)) = 0$, for every sequence of sets $(A_n)_{n\in\mathbb{N}^*} \subset \mathcal{C}$, with $A_n \nearrow A \in \mathcal{C}$;

(IV) μ is said to be a *multisubmeasure* if it is monotone and

$\mu(A \cup B) \subseteq \mu(A) \dot{+} \mu(B)$, for every $A, B \in \mathcal{C}$, with $A \cap B = \emptyset$ (or, equivalently, for every $A, B \in \mathcal{C}$);

(V) μ is said to be a *multimeasure* if $\mu(A \cup B) = \mu(A) \dot{+} \mu(B)$, for every $A, B \in \mathcal{C}$, with $A \cap B = \emptyset$;

(VI) μ is said to be a *fuzzy multivalued set function* if μ is monotone, increasing convergent and decreasing convergent;

(VII) μ is said to be a *fuzzy multisubmeasure* (a *fuzzy multimeasure,* respectively) if μ is a fuzzy multivalued set function and it also is a multisubmeasure (multimeasure, respectively);

(VIII) A *multisubmeasure* or *multimeasure* or *fuzzy multivalued set function* defined on $\mathcal{B}_0$ (or, respectively, $\mathcal{B}$) will be shortly called a *Baire* (or, respectively, *Borel*) one.

Remark 4.2.2

(i) Evidently, any monotone multimeasure is, particularly, a multisubmeasure.

(ii) Any multisubmeasure on a ring $\mathcal{C}$ is o-continuous if and only if it is a fuzzy multivalued set function. Indeed, according to [6], if μ is o-continuous, then μ is increasing convergent and decreasing convergent, hence μ is a fuzzy multisubmeasure.

Conversely, if μ is a fuzzy multisubmeasure, then μ is decreasing convergent, hence it is o-continuous.

(iii) If $\mathcal{C}$ is a σ-ring and $\mu : \mathcal{C} \to \mathcal{P}_f(X)$ is a decreasing convergent, monotone multivalued set function, with $\mu(\emptyset) = \{0\}$, then μ is exhaustive. Indeed, if $(A_n)_{n\in\mathbb{N}^*} \subset \mathcal{C}$ is a pairwise disjoint sequence of sets, let $(B_n)_{n\in\mathbb{N}^*} \subset \mathcal{C}$, defined by $B_n = \bigcup\limits_{k=n+1}^{\infty} A_k$, for every $n \in \mathbb{N}^*$. Then $B_n \searrow \emptyset$, and, since μ is decreasing convergent, then

$$\lim_{n\to\infty} |\mu(B_n)| = \lim_{n\to\infty} h(\mu(B_n), \{0\}) = 0.$$

Because μ is monotone and $A_n \subset B_{n-1}$, then $e(\mu(A_n), \mu(B_{n-1})) = 0$, hence

$$|\mu(A_n)| \leq |\mu(B_{n-1})|, \text{ for every } n \geq 2.$$

Consequently, $\lim\limits_{n\to\infty} |\mu(A_n)| = 0$ and, therefore, μ is exhaustive.

(iv) Any fuzzy multivalued set function on a σ-ring is exhaustive.

Example 4.2.3

(i) If $\nu : \mathcal{C} \to \mathbb{R}_+$ is an o-continuous submeasure (finitely additive measure, respectively), then the multivalued set function $\mu : \mathcal{C} \to \mathcal{P}_f(\mathbb{R})$, defined by

$$\mu(A) = [0, \nu(A)], \text{ for every } A \in \mathcal{C},$$

is a fuzzy multisubmeasure (multimeasure, respectively), called *the multisubmeasure (multimeasure,* respectively*) induced by* ν.

(ii) If $m_1, m_2 : \mathcal{C} \to \mathbb{R}_+$, m_1 is an o-continuous finitely additive measure and m_2 is an o-continuous submeasure (finitely additive measure, respectively), then the multivalued set function $\mu : \mathcal{C} \to \mathcal{P}_f(\mathbb{R})$, defined by

$$\mu(A) = [-m_1(A), m_2(A)], \text{ for every } A \in \mathcal{C},$$

is a fuzzy multisubmeasure (multimeasure, respectively).

In [10] we have introduced for multivalued set functions the following notions: of an atom, non-atomicity and the Darboux property. Refer also to Section 3.2 for the definitions of an atom, respectively, a non-atomic set multifunction.

Definition 4.2.4 Let $\mu : \mathcal{C} \to \mathcal{P}_f(X)$ be a multivalued set function, with $\mu(\emptyset) = \{0\}$. We say that μ has *the Darboux property* if for every $A \in \mathcal{C}$, with $\mu(A) \supsetneq \{0\}$ and every $p \in (0, 1)$, there exists a set $B \in \mathcal{C}$ such that $B \subset A$ and $\mu(B) = p\,\mu(A)$.

We recall now different types of regularity we defined and studied in [6, 7] and [8] for multisubmeasures with respect to the Hausdorff topology (see also Chapter 2).

Definition 4.2.5 Let $A \in \mathcal{C}$ be an arbitrary set and $\mu : \mathcal{C} \longrightarrow \mathcal{P}_f(X)$ a multisubmeasure.

(I) A is said to be with respect to μ if for every $\varepsilon > 0$, there exist a compact set $K \subset A$, $K \in \mathcal{C}$ and an open set $D \supset A$, $D \in \mathcal{C}$ such that $|\mu(B)| < \varepsilon$, for every $B \in \mathcal{C}$, $B \subset D \backslash K$;

(II) A is said to be *R_l'-regular* with respect to μ if for every $\varepsilon > 0$, there exists a compact set $K \subset A$, $K \in \mathcal{C}$ such that $|\mu(B)| < \varepsilon$, for every $B \in \mathcal{C}$, $B \subset A \backslash K$;

(III) A is said to be *R_r'-regular* with respect to μ if for every $\varepsilon > 0$, there exists an open set $D \supset A$, $D \in \mathcal{C}$ such that $|\mu(B)| < \varepsilon$, for every $B \in \mathcal{C}$, $B \subset D \backslash A$;

(IV) A is said to be *R-regular* with respect to μ if for every $\varepsilon > 0$, there exist a compact set $K \subset A$, $K \in \mathcal{C}$ and an open set $D \supset A$, $D \in \mathcal{C}$ such that $h(\mu(A), \mu(B)) < \varepsilon$, for every $B \in \mathcal{C}$, $K \subset B \subset D$;

(V) A is said to be *R_l-regular* with respect to μ if for every $\varepsilon > 0$, there exists a compact set $K \subset A$, $K \in \mathcal{C}$ such that $h(\mu(A), \mu(B)) < \varepsilon$, for every $B \in \mathcal{C}$, $K \subset B \subset A$;

(VI) A is said to be *R_r-regular* with respect to μ if for every $\varepsilon > 0$, there exists an open set $D \supset A$, $D \in \mathcal{C}$ such that $h(\mu(A), \mu(B)) < \varepsilon$, for every $B \in \mathcal{C}$, $A \subset B \subset D$;

(VII) μ is said to be *R'-regular* (*R_l'-regular*, *R_r'-regular*, respectively) if every $A \in \mathcal{C}$ is a R'-regular (R_l'-regular, R_r'-regular, respectively) set with respect to μ;

(VIII) μ is said to be *R-regular* (*R_l-regular*, *R_r-regular*, respectively) if every $A \in \mathcal{C}$ is a R-regular (R_l-regular, R_r-regular, respectively) set with respect to μ.

Remark 4.2.6

(i) If μ is a R_l'-regular multisubmeasure on a ring $\mathcal{C}$, then μ is a fuzzy multivalued set function. Indeed, if μ is a R_l'-regular multisubmeasure, then, according to [7] or [8], it is also o-continuous, hence by Remark 4.2.2(ii), μ is a fuzzy multivalued set function.

(ii) If $\mathcal{C} = \mathcal{B}_0$, then a multisubmeasure μ is R_l'-regular if and only if it is a fuzzy multivalued set function. Indeed, according to [7], a multisubmeasure $\mu : \mathcal{B}_0 \to \mathcal{P}_f(X)$ is R_l'-regular if and only if it is o-continuous. By Remark 4.2.2(ii), we immediately get the conclusion.

(iii) For a multisubmeasure defined on $\mathcal{B}_0$ or $\mathcal{B}$, R'-regularity is equivalent to R_l'-regularity (see [6]). So, if $\mu : \mathcal{B}_0 \to \mathcal{P}_f(X)$ is a multisubmeasure, then μ is R'-regular if and only if it is a fuzzy multivalued set function.

4.3 Non-atomic multisubmeasures

Theorem 4.3.1 *Let be $A \in \mathcal{B}$ with $\mu(A) \supsetneq \{0\}$ and $\mu : \mathcal{B} \to \mathcal{P}_f(X)$ a R_l'-regular multisubmeasure.*

(i) A is an atom of μ if and only if

$$(*)\ \exists! a \in A \text{ so that } \mu(A \backslash \{a\}) = \{0\};$$

(ii) If

$$(**) \qquad \text{for every } t \in T, \text{ there exists } A_t \in \mathcal{B} \text{ so that } t \in A_t \text{ and } \mu(A_t) = \{0\},$$

then μ is non-atomic.

Proof The "Only if part". Let $A \in \mathcal{B}$ be an atom of μ and $\mathcal{K}_A = \{K \subseteq A;\ K$ is a compact set and $\mu(A \backslash K) = \{0\}\} \subset \mathcal{B}$. One can easily check that $\mathcal{K}_A$ is nonvoid since μ is R_l'-regular. Indeed, if we suppose by the contrary that $\mathcal{K}_A = \emptyset$, then for every compact set $K \subseteq A$, we have $\mu(A \backslash K) = \{0\}$. Since A is an atom of μ, then $\mu(K) = \{0\}$. Then, by the R_l'-regularity of μ we get that $\mu(A) = \{0\}$, which is a contradiction.

(i) We now prove that every $K \in \mathcal{K}_A$ is an atom of μ. Indeed, if $K \in \mathcal{K}_A$, then

$$\{0\} \subsetneq \mu(A) \subseteq \mu(A \backslash K) \dot{+} \mu(K) = \mu(K),$$

hence $\mu(K) \supsetneq \{0\}$. Also, for every $B \in \mathcal{B}$, with $B \subseteq K$, since $K \subseteq A$ and A is an atom of μ, we get that $\mu(B) = \{0\}$ or $\mu(A \backslash B) = \{0\}$. If $\mu(A \backslash B) = \{0\}$, then

$$\{0\} \subseteq \mu(K \backslash B) \subseteq \mu(A \backslash B) = \{0\},$$

hence $\mu(K \backslash B) = \{0\}$.

We prove now that $K_1 \cap K_2 \in \mathcal{K}_A$, for every $K_1, K_2 \in \mathcal{K}_A$. Indeed, if $K_1, K_2 \in \mathcal{K}_A$, then $K_1 \cap K_2$ is a compact set of T and

$$\{0\} \subseteq \mu(A \backslash (K_1 \cap K_2)) \subseteq \mu(A \backslash K_1) \dot{+} \mu(A \backslash K_2) = \{0\}.$$

Consequently, $\mu(A \backslash (K_1 \cap K_2)) = \{0\}$.

We prove that $\bigcap_{K \in \mathcal{K}_A} K$, denoted by K_0, is a nonvoid set. Suppose that, on the contrary, $K_0 = \emptyset$. Then there exist $K_1, K_2, \ldots, K_{n_0} \in \mathcal{K}_A$ so that $\bigcap_{i=1}^{n_0} K_i = \emptyset$, hence $\mu(\bigcap_{i=1}^{n_0} K_i) = \{0\}$. But $\bigcap_{i=1}^{n_0} K_i \in \mathcal{K}_A$, which implies that $\mu(\bigcap_{i=1}^{n_0} K_i) \supsetneq \{0\}$, a contradiction.

Now, we prove that $K_0 \in \mathcal{K}_A$. Obviously, K_0 is a compact set. Let be $K \in \mathcal{K}_A$. Then $\mu(A \backslash K) = \{0\}$.

If $K = K_0$, then $K_0 \in \mathcal{K}_A$.

If $K \neq K_0$, then $K_0 \subsetneq K$.

Because

$$\{0\} \subseteq \mu(A \backslash K_0) \subseteq \mu(A \backslash K) \overset{\bullet}{+} \mu(K \backslash K_0) = \mu(K \backslash K_0),$$

it is sufficient to demonstrate that $\mu(K \backslash K_0) = \{0\}$.

Suppose that, on the contrary, $\mu(K \backslash K_0) \supsetneq \{0\}$. Let $B \in \mathcal{B}$, with $B \subseteq K \backslash K_0$. Then $B \subseteq K$ and, since K is an atom of μ, then $\mu(B) = \{0\}$ or $\mu(K \backslash B) = \{0\}$. If $\mu(K \backslash B) = \{0\}$, then $\mu((K \backslash K_0) \backslash B) = \{0\}$, hence $K \backslash K_0$ is an atom of μ. Because A is an atom of μ and $\mu(K \backslash K_0) \supsetneq \{0\}$, then $\mu(A \backslash (K \backslash K_0)) = \{0\}$.

Consequently, $\mathcal{K}_A = \{B \subseteq A;\ B$ is a compact set and $\mu(A \backslash B) = \{0\}\}$ and $\mathcal{K}_{K \backslash K_0} = \{C \subseteq K \backslash K_0;\ C$ is a compact set and $\mu((K \backslash K_0) \backslash C) = \{0\}\}$.

Let be $C \in \mathcal{K}_{K \backslash K_0}$. Then $\mu((K \backslash K_0) \backslash C) = \{0\}$ and, since $\mu(A \backslash (K \backslash K_0)) = \{0\}$, we get that

$$\{0\} \subseteq \mu(A \backslash C) \subseteq \mu(A \backslash (K \backslash K_0)) \overset{\bullet}{+} \mu((K \backslash K_0) \backslash C) = \{0\}.$$

hence $\mu(A \backslash C) = \{0\}$, which implies that $C \in \mathcal{K}_A$. Therefore, $K_0 \subseteq C$, but $C \subseteq K \backslash K_0$, a contradiction. Consequently, $\mu(K \backslash K_0) = \{0\}$.

We now prove that the set K_0 is a singleton $\{a\}$. Suppose that, on the contrary, there exist $a, b \in A$, with $a \neq b$ and $K_0 \supseteq \{a, b\}$.

Since T is a locally compact Hausdorff space, there exists an open neighbourhood V of a so that $b \notin \overline{V}$. Obviously, $K_0 = (K_0 \backslash V) \cup (K_0 \cap \overline{V})$ and $K_0 \backslash V$, $K_0 \cap \overline{V}$ are nonvoid, compact subsets of A.

We prove that $K_0 \backslash V \in \mathcal{K}_A$ or $K_0 \cap \overline{V} \in \mathcal{K}_A$. Indeed, if $K_0 \backslash V \notin \mathcal{K}_A$ and $K_0 \cap \overline{V} \notin \mathcal{K}_A$, then $\mu(A \backslash (K_0 \backslash (V)) \supsetneq \{0\}$ and $\mu(A \backslash (K_0 \cap \overline{V})) \supsetneq \{0\}$. Since A is an atom of μ, $A \backslash (K_0 \backslash V) \subseteq A$ and $A \backslash (K_0 \cap \overline{V}) \subseteq A$, then $\mu(K_0 \backslash V) = \{0\}$ and $\mu(K_0 \cap \overline{V}) = \{0\}$. Hence, $\mu(K_0) = \{0\}$, which implies that

$$\{0\} \subsetneq \mu(A) \subseteq \mu(A \backslash K_0) \overset{\bullet}{+} \mu(K_0) = \{0\},$$

a contradiction. Then $K_0 \backslash V \in \mathcal{K}_A$ or $K_0 \cap \overline{V} \in \mathcal{K}_A$. Because $K_0 \subseteq K$, for every $K \in \mathcal{K}_A$, we get that $K_0 \subseteq K_0 \backslash V$ or $K_0 \subseteq K_0 \cap \overline{V}$, which is impossible. The uniqueness is immediate. So, $\exists! a \in A$ so that $\mu(A \backslash \{a\}) = \{0\}$, as claimed.

We note that we also have $\mu(A) = \mu(\{a\})$.

The "If part".

We consider $A \in \mathcal{B}$, with $\mu(A) \supsetneq \{0\}$ having the property $(*)$ and let be $B \in \mathcal{B}$, with $B \subset A$. If $a \notin B$, then $B \subset A \backslash \{a\}$. Because $\mu(A \backslash \{a\}) = \{0\}$, then $\mu(B) = \{0\}$. If $a \in B$, then $A \backslash B \subset A \backslash \{a\}$, hence $\mu(A \backslash B) = \{0\}$. Consequently, A is an atom of μ.

(ii) Suppose that, on the contrary, there exists an atom $A_0 \in \mathcal{B}$ of μ. Then, by (ii) we get that $\exists! a \in A_0$ so that $\mu(A_0 \backslash \{a\}) = \{0\}$. On the other hand, by $(**)$ we have that for $a \in A_0$, there exists $\widetilde{A} \in \mathcal{B}$ such that $a \in \widetilde{A}$ and $\mu(\widetilde{A}) = \{0\}$. Because $A_0 \subseteq (A_0 \backslash \{a\}) \cup \widetilde{A}$, we get that

$$\{0\} \subsetneq \mu(A_0) \subseteq \mu(A_0 \backslash \{a\}) \stackrel{\bullet}{+} \mu(\widetilde{A}) = \{0\},$$

a contradiction. Consequently, μ is non-atomic.

□

Remark 4.3.2 The condition (∗∗) may be replaced by the condition
(∗ ∗ ∗) for every $t \in T$, $\mu(\{t\}) = \{0\}$.
Indeed, since T is a locally compact Hausdorff space, then $\{t\}$ is a compact set, hence $\{t\} \in \mathcal{B}$.

In the sequel, we establish a converse of Theorem 4.3.1(iii):

Theorem 4.3.3 *If $\mu : \mathcal{B} \to \mathcal{P}_f(X)$ is a Borel non-atomic multisubmeasure, then* (∗ ∗ ∗) *for every $t \in T$, $\mu(\{t\}) = \{0\}$.*

Proof Suppose that, on the contrary, there exists $t_0 \in T$ so that $\mu(\{t_0\}) \supsetneq \{0\}$. Because μ is non-atomic, there is a set $B \in \mathcal{B}$ such that $B \subseteq \{t_0\}$, $\mu(B) \supsetneq \{0\}$ and $\mu(\{t_0\} \backslash B) \supsetneq \{0\}$. Consequently, $B = \emptyset$ or $B = \{t_0\}$, which is false. □

Using Theorem 4.3.1(iii), Remark 4.3.2 and Theorem 4.3.3, we immediately obtain the following necessary and sufficient condition for the non-atomicity of a R_l'-regular Borel multisubmeasure:

Corollary 4.3.4 *Let $\mu : \mathcal{B} \to \mathcal{P}_f(X)$ be a R_l'-regular Borel multisubmeasure. Then μ is non-atomic if and only if for every $t \in T$, $\mu(\{t\}) = \{0\}$.*

Further, we establish a condition which is weaker than (∗∗), for the non-atomicity of a R-regular multisubmeasure on $\mathcal{B}_0$ (or $\mathcal{B}$):

Theorem 4.3.5

(i) Let be $\mathcal{C} = \mathcal{B}_0$ (or $\mathcal{B}$) and $\mu : \mathcal{C} \to \mathcal{P}_f(X)$ a multisubmeasure. If μ is and if it has the property

$$(o) \qquad \text{for every } B \in \mathcal{C}, \text{ with } \mu(B) \supsetneq \{0\} \text{ and every } t \in T,$$
$$\text{there exists } A_t \in \mathcal{C} \text{ so that } t \in A_t \text{ and } e(\mu(B), \mu(A_t)) > 0,$$

then μ is non-atomic;

(ii) Let $\mu : \mathcal{B}_0 \to \mathcal{P}_f(X)$ be a fuzzy multisubmeasure having the property (o). Then μ is non-atomic.

Proof

(i) Suppose that, on the contrary, there exists an atom $B \in \mathcal{C}$ of μ.
Because μ is R-regular then, according to [6], it is R_l-regular. Then there exists a compact set $K \in \mathcal{C}$ so that $K \subseteq B$ and $h(\mu(B), \mu(K)) < |\mu(B)|$.
We observe that $\mu(K) \supsetneq \{0\}$. Indeed, if $\mu(K) = \{0\}$, then $|\mu(B)| < |\mu(B)|$, which is false.

According to (o), for every $t \in K$, there exists $A_t \in \mathcal{C}$ so that $t \in A_t$ and $e(\mu(B), \mu(A_t)) > 0$.

Because μ is R-regular then, according to [6], it is R_r-regular. Then, for every $t \in K$, for A_t there exists an open set $D_t \in \mathcal{C}$ so that $A_t \subseteq D_t$ and

$$e(\mu(D_t), \mu(A_t)) \leq h(\mu(D_t), \mu(A_t)) < e(\mu(B), \mu(A_t)).$$

Since $t \in A_t$ and $A_t \subseteq D_t$, then $K \subseteq \bigcup_{t\in K} D_t$. Consequently, there exists $p \in \mathbb{N}^*$ such that $K \subseteq \bigcup_{i=1}^{p} D_{t_i}$, with $t_i \in K$, for every $i = \overline{1, p}$. This implies

$$\{0\} \subsetneq \mu(K) = \mu(\bigcup_{i=1}^{p} (D_{t_i} \cap K)) \subseteq \mu(D_{t_1} \cap K) \dot{+} \mu(D_{t_2} \cap K) \dot{+} \cdots \dot{+} \mu(D_{t_p} \cap K).$$

Then there is $s = \overline{1, p}$ such that $\mu(D_{t_s} \cap K) \supsetneq \{0\}$. Consequently,

$$\{0\} \subsetneq \mu(D_{t_s} \cap K) \subseteq \mu(K) \subseteq \mu(B).$$

Obviously, we also have $\mu(D_{t_s}) \supsetneq \{0\}$.

Since B is an atom of μ, $\mu(B) \supsetneq \{0\}$ and $\mu(D_{t_s}) \supsetneq \{0\}$, we get that $\mu(B \backslash D_{t_s}) = \{0\}$, hence

$$\mu(B \cap D_{t_s}) \subseteq \mu(B) \subseteq \mu(B \backslash D_{t_s}) \dot{+} \mu(B \cap D_{t_s}) = \mu(B \cap D_{t_s}),$$

which implies $\mu(B) = \mu(D_{t_s} \cap B)$.

On the other hand, because $e(\mu(D_{t_s}), \mu(A_{t_s})) < e(\mu(B), \mu(A_{t_s}))$, then

$$\begin{aligned} e(\mu(B \cap D_{t_s}), \mu(A_{t_s})) &\leq e(\mu(B \cap D_{t_s}), \mu(D_{t_s})) + e(\mu(D_{t_s}), \mu(A_{t_s})) \\ &= e(\mu(D_{t_s}), \mu(A_{t_s})) < e(\mu(B), \mu(A_{t_s})). \end{aligned}$$

But $\mu(B) = \mu(D_{t_s} \cap B)$, a contradiction. So, μ is non-atomic.

(ii) It easily results according to (i) and Remark 4.2.6(ii) and (iii), also taking into account that R'-regularity implies R-regularity (see [6]).

□

We shall obtain in Chapter 9 a generalization of the previous theorem.

Now, we prove a converse of Theorem 4.3.5(i) for the particular case when the multisubmeasure is a monotone multimeasure:

Theorem 4.3.6 *If $\mathcal{C} = \mathcal{B}_0$ (or $\mathcal{B}$) and $\mu : \mathcal{C} \to \mathcal{P}_{bfc}(X)$ is a non-atomic monotone multimeasure, then μ has the property (o).*

Proof Suppose that, on the contrary, there exist $B_0 \in \mathcal{C}$, with $\mu(B_0) \supsetneq \{0\}$ and $t_0 \in B_0$, so that for every $A \in \mathcal{C}$, with $t_0 \in A$, we have $e(\mu(B_0), \mu(A_t)) = 0$. Then $\mu(B_0) \subseteq \mu(A_t)$.

Because μ is non-atomic, there exists $S_0 \in \mathcal{C}$ so that $S_0 \subseteq B_0$, $\mu(S_0) \not\supseteq \{0\}$ and $\mu(B_0 \backslash S_0) \not\supseteq \{0\}$.

(a) If $t_0 \in S_0$, then $\mu(B_0) \subseteq \mu(S_0)$. Since $S_0 \subseteq B_0$, we get that $\mu(B_0) = \mu(S_0)$. Then $\mu(S_0) = \mu(S_0) \overset{\bullet}{+} \mu(B_0 \backslash S_0)$. Using the cancelation law in $\mathcal{P}_{bfc}(X)$, we have $\mu(B_0 \backslash S_0) = \{0\}$, a contradiction.

(b) If $t_0 \in B_0 \backslash S_0$, then $\mu(B_0) \subseteq \mu(B_0 \backslash S_0)$. As before, we immediately get that $\mu(S_0) = \{0\}$, a contradiction.

□

Corollary 4.3.7 *Let be $\mathcal{C} = \mathcal{B}_0$ (or $\mathcal{B}$) and $\mu : \mathcal{C} \to \mathcal{P}_{bfc}(X)$ a R-regular monotone multimeasure. Then μ is non-atomic if and only if μ has the property (o).*

Now, it is easy to observe that, if $\mathcal{C}_1 \subset \mathcal{C}_2$ are two rings and if $\mu : \mathcal{C}_2 \to \mathcal{P}_f(X)$ is a non-atomic multisubmeasure, $\mu_{/\mathcal{C}_1}$ may not be non-atomic. Although, we establish the following:

Theorem 4.3.8 *Let $\mu : \mathcal{B} \to \mathcal{P}_f(X)$ be a R'-regular, non-atomic Borel multisubmeasure. Then $\mu/_{\mathcal{B}_0} : \mathcal{B}_0 \to \mathcal{P}_f(X)$ is also a R'-regular, non-atomic Baire multisubmeasure.*

Proof Because μ is R'-regular on $\mathcal{B}$, then, according to [7], it is o-continuous on $\mathcal{B}$. Obviously, it is also o-continuous on $\mathcal{B}_0$. By [7], this is equivalent to the R'-regularity of μ on $\mathcal{B}_0$. Also, μ is R-regular on $\mathcal{B}_0$.

We prove now that μ is non-atomic on $\mathcal{B}_0$.

Because μ is non-atomic on $\mathcal{B}$, then, by Theorem 4.3.3, we get that for every $t \in T$, $\mu(\{t\}) = \{0\}$.

Since μ is R'-regular on $\mathcal{B}$, then, according to [6], it is also R-regular on $\mathcal{B}$, so, by Gavriluţ [6], it is R_r-regular on $\mathcal{B}$. Then for every $n \in \mathbb{N}^*$, there exists an open set $D_n \in \mathcal{B}$ so that $\{t\} \subseteq D_n$ and $h(\mu(\{t\}), \mu(D_n)) < \frac{1}{n}$. Because $\mu(\{t\}) = \{0\}$, this yields $|\mu(D_n)| < \frac{1}{n}$, for every $n \in \mathbb{N}^*$.

On the other hand, because T is a locally compact Hausdorff space, $\{t\}$ is a compact set, $\{t\} \subseteq D_n$ and D_n is an open set, for every $n \in \mathbb{N}^*$, by Dinculeanu [2] we get that for every $n \in \mathbb{N}^*$ there exists an open set $\widetilde{D}_n \in \mathcal{B}_0$ such that $\{t\} \subseteq \widetilde{D}_n \subseteq D_n$. Because $|\mu(D_n)| < \frac{1}{n}$, we also have $|\mu(\widetilde{D}_n)| < \frac{1}{n}$, for every $n \in \mathbb{N}^*$.

Let $M = \bigcap_{n=1}^{\infty} \widetilde{D}_n$. We denote $E_n = \bigcap_{i=1}^{n} \widetilde{D}_i$, for every $n \in \mathbb{N}^*$. Then $E_n \searrow M$, $(E_n)_n \subset \mathcal{B}_0$, $M \in \mathcal{B}_0$ and $t \in M$.

Since $E_n \subseteq \widetilde{D}_n$, then $|\mu(E_n)| < \frac{1}{n}$, for every $n \in \mathbb{N}^*$, which implies

$$\begin{aligned} |\mu(M)| &\leq |\mu(E_n)| + h(\mu(E_n), \mu(M)) \\ &< \frac{1}{n} + h(\mu(E_n), \mu(M)), \text{ for every } n \in \mathbb{N}^*. \end{aligned}$$

Now, because μ is o-continuous, then by Gavriluţ [6] it is decreasing convergent, hence $\lim_{n \to \infty} h(\mu(E_n), \mu(M)) = 0$. The last inequalities imply then that $|\mu(M)| = 0$, hence $\mu(M) = \{0\}$.

Consequently, for every $t \in T$, there exists $M \in \mathcal{B}_0$ so that $t \in M$ and $\mu(M) = \{0\}$, hence, by Theorem 4.3.5(i), $\mu/_{\mathcal{B}_0}$ is non-atomic. □

Corollary 4.3.9

(i) *If X is a Banach space and if $\nu : \mathcal{B}_0 \to \mathcal{P}_{bf}(X)$ is a R'-regular Baire multisubmeasure, which has atoms, then ν uniquely extends to a R'-regular Borel multisubmeasure $\mu : \mathcal{B} \to \mathcal{P}_{bf}(X)$ which also has atoms.*

(ii) *Let X be a Banach space and $\nu : \mathcal{B}_0 \to \mathcal{P}_{bf}(X)$ a fuzzy Baire multisubmeasure which has atoms. Then ν uniquely extends to a fuzzy Borel multisubmeasure $\mu : \mathcal{B} \to \mathcal{P}_{bf}(X)$ which also has atoms.*

Proof

(i) Let $\nu : \mathcal{B}_0 \to \mathcal{P}_f(X)$ be a R'-regular multisubmeasure, which has atoms. By Remark 4.2.6(iii), R'-regularity is equivalent to R'_l-regularity. So, ν is R'_l-regular. According to [6], then $\nu : \mathcal{B}_0 \to \mathcal{P}_{bf}(X)$.

On the other hand, by Gavriluţ [9] (or [8]), we know that ν uniquely extends to a R'-regular multisubmeasure $\mu : \mathcal{B} \to \mathcal{P}_{bf}(X)$ so that $\mu_{/\mathcal{B}_0} = \nu$. We observe that μ has atoms. Indeed, let us suppose that, on the contrary, μ is non-atomic. So, $\mu : \mathcal{B} \to \mathcal{P}_{bf}(X)$ is a non-atomic, R'-regular multisubmeasure, hence, by Theorem 4.3.8 we get then that $\mu_{/\mathcal{B}_0} = \nu$ is also non-atomic, a contradiction.

(ii) If $\nu : \mathcal{B}_0 \to \mathcal{P}_f(X)$ is a fuzzy multisubmeasure, then, according to Remark 4.2.6(ii), it is R'_l-regular. Since Remark 4.2.6(iii), we get that ν is R'-regular, hence by (i) it uniquely extends to a R'-regular multisubmeasure $\mu : \mathcal{B} \to \mathcal{P}_{bf}(X)$. Because R'-regularity implies o-continuity (see [7]), then μ is o-continuous. Consequently, by Remark 4.2.2(ii), μ is a fuzzy multisubmeasure. □

4.4 The Darboux property

In this subsection, we establish several results concerning Borel/Baire multisubmeasures and monotone multimeasures having the Darboux property. We also point out connections between non-atomicity and the Darboux property.

4.4.1 The Darboux property for multisubmeasures

Theorem 4.4.1 *If $\mu : \mathcal{B} \to \mathcal{P}_f(X)$ is a Borel multisubmeasure, with $|\mu(A)| < \infty$, for every $A \in \mathcal{B}$, having the Darboux property, then for every $t \in T$, $\mu(\{t\}) = \{0\}$.*

Proof The statement is an immediate consequence of Corollary 4.3.4 and of [10] (for any multisubmeasure defined on a ring $\mathcal{C}$, satisfying the condition $|\mu(A)| < \infty$, for every $A \in \mathcal{C}$, the Darboux property always implies non-atomicity). Of course, if $\mu : \mathcal{C} \to \mathcal{P}_{bf}(X)$ is an arbitrary multisubmeasure, then for every $A \in \mathcal{C}$, $\mu(A)$ is a bounded set, so $|\mu(A)| < \infty$. □

The converse of Theorem 4.4.1 is not valid in general, as we easily see from the following example:

Example 4.4.2 Gavriluţ [10] Let T be a compact space and $\mu : \mathcal{B} \to \mathcal{P}_{kc}(\mathbb{R})$ be the Borel multisubmeasure defined by:

$$\mu(A) = \begin{cases} [-m(A), m(A)], & \text{if } m(A) \leq 1 \\ [-m(A), 1], & \text{if } m(A) > 1 \end{cases}, \text{ for every } A \in \mathcal{B},$$

where $m : \mathcal{B} \to \mathbb{R}_+$ is a bounded, finitely additive set function having the Darboux property.

Then μ is non-atomic, hence by Corollary 4.3.4, $\mu(\{t\}) = \{0\}$, for every $t \in T$, but μ does not have the Darboux property (see [10] for details).

4.4.2 The Darboux property for multimeasures

Remark 4.4.3 If $\mu : \mathcal{B} \to \mathcal{P}_f(X)$ is a Borel monotone multimeasure, with $|\mu(A)| < \infty$, for every $A \in \mathcal{B}$, also having the Darboux property, then for every $t \in T$, $\mu(\{t\}) = \{0\}$. Indeed, we apply Theorem 4.4.1 for monotone multimeasures, instead of multisubmeasures.

Theorem 4.4.4 *Let be $\mathcal{C} = \mathcal{B}_0$ (or $\mathcal{B}$). If $\mu : \mathcal{C} \to \mathcal{P}_{bfc}(X)$ is a monotone, decreasing convergent (or, moreover, a fuzzy) multimeasure, having the Darboux property, then*

(∗∗) *for every $t \in T$, there exists $A_t \in \mathcal{C}$ so that $t \in A_t$ and $\mu(A_t) = \{0\}$.*

Proof Let $t \in T$ be arbitrary, but fixed. We first prove that there exists $A_t \in \mathcal{C}$ so that $t \in A_t$.

(a) If $\mathcal{C} = \mathcal{B}$, then $A_t = \{t\}$.
(b) If $\mathcal{C} = \mathcal{B}_0$, since T is a locally compact Hausdorff space, then for t there exists a compact neighbourhood V_t, for which there is a relatively compact, open set D_t so that $V_t \subseteq D_t$. By Belley, and Morales [1] we know that there exists $A_t \in \mathcal{B}_0$ so that $V_t \subseteq A_t \subseteq D_t$. Evidently, $t \in A_t$.

Let now $A = A_t \in \mathcal{C}$ be so that $t \in A$.

If $\mu(A) = \{0\}$, the proof is finished.

If $\mu(A) \supsetneq \{0\}$, since μ has the Darboux property, there exists a set $A_1 \in \mathcal{C}$ so that $A_1 \subseteq A$ and $\mu(A_1) = \frac{1}{2}\mu(A)$. Then

$$\mu(A) = \frac{1}{2}\mu(A) \overset{\bullet}{+} \mu(A\backslash A_1) = \frac{1}{2}\mu(A) \overset{\bullet}{+} \frac{1}{2}\mu(A).$$

Using the cancelation law in $\mathcal{P}_{bfc}(X)$, we get that $\mu(A\backslash A_1) = \frac{1}{2}\mu(A)$.

Let $B_1 = A_1$ or $B_1 = A\backslash A_1$ be so that $t \in B_1$. Obviously, $\mu(B_1) = \frac{1}{2}\mu(A)$.

We construct by induction a decreasing sequence of sets $(B_n)_n \subset \mathcal{C}$ such that $\mu(B_n) = \frac{1}{2^n}\mu(A)$, $B_n \subseteq A$ and $t \in B_n$, for every $n \in \mathbb{N}^*$.

We suppose that we have already constructed $B_1, B_2, \dots B_n$. Let us obtain B_{n+1} :

By the Darboux property, there exists $A_{n+1} \in \mathcal{C}$ so that $A_{n+1} \subseteq B_n$ and $\mu(A_{n+1}) = \frac{1}{2}\mu(B_n) = \cdots = \frac{1}{2^{n+1}}\mu(A)$. Then $\mu(B_n\backslash A_{n+1}) = \frac{1}{2}\mu(B_n) = \frac{1}{2^{n+1}}\mu(A)$.

Let $B_{n+1} = A_{n+1}$ or $B_{n+1} = B_n\backslash A_{n+1}$ be so that $t \in B_{n+1}$. Then $\mu(B_{n+1}) = \frac{1}{2^{n+1}}\mu(A)$.

We consider $B = \bigcap_{n=1}^{\infty} B_n$. Obviously, $t \in B$, $B \in \mathcal{C}$ and, for every $n \in \mathbb{N}^*$, we have

$$|\mu(B)| \leq h(\mu(B_n), \mu(B)) + |\mu(B_n)| = h(\mu(B_n), \mu(B)) + \frac{1}{2^n}|\mu(A)|.$$

Because $\mu : \mathcal{C} \to \mathcal{P}_{bfc}(X)$, then $\mu(A)$ is a bounded set, that is, there exists $M > 0$ so that $|\mu(A)| \leq M$. Also, because μ is decreasing convergent, $\lim_{n\to\infty} h(\mu(B_n), \mu(B)) = 0$. Therefore, $|\mu(B)| = 0$, hence $\mu(B) = \{0\}$.

So, for every $t \in T$, there exists $B \in \mathcal{C}$ so that $t \in B$ and $\mu(B) = \{0\}$. □

Remark 4.4.5 Of course, if in the above theorem, $\mathcal{C} = \mathcal{B}$, since $\mu : \mathcal{B} \to \mathcal{P}_{bfc}(X)$, then $|\mu(A)| < \infty$, for every $A \in \mathcal{C}$. Because μ is a monotone multimeasure having the Darboux property, then our result follows immediately. If $\mathcal{C} = \mathcal{B}_0$, we observe that, in this case, $\mu : \mathcal{B}_0 \to \mathcal{P}_{bfc}(X)$ is a monotone multimeasure and it is also non-atomic (because $|\mu(A)| < \infty$, for every $A \in \mathcal{C}$ and μ has the Darboux property). By Theorem 4.3.6, we observe that μ has the property (o). But $(**)$ always implies (o), hence our theorem proves that if, moreover, μ is decreasing convergent, then μ has, moreover, the property $(**)$, not only (o).

In [10] we have proved that, for multisubmeasures defined on a ring $\mathcal{C}$, with $|\mu(A)| < \infty$, for every $A \in \mathcal{C}$, the Darboux property always implies non-atomicity. In the sequel, we establish that, for R'-regular, monotone multimeasures defined on $\mathcal{B}_0$ or $\mathcal{B}$ (or, for fuzzy multimeasures defined on $\mathcal{B}_0$), the property $(**)$ stands between Darboux property and non-atomicity.

Corollary 4.4.6

(i) Let be $\mathcal{C} = \mathcal{B}_0$ (or $\mathcal{B}$). If $\mu : \mathcal{C} \to \mathcal{P}_{bfc}(X)$ is a R'-regular monotone multimeasure, then

(oo) "Darboux property" → "$(**)$" → "non-atomicity";

(ii) If $\mu : \mathcal{B}_0 \to \mathcal{P}_{bfc}(X)$ is a fuzzy Baire multimeasure, then μ has the property (oo).

Proof

(i) Since μ is R'-regular, then it is R'_l-regular and, consequently, exhaustive, so, by Gavriluţ [6], $|\mu(A)| < \infty$, for every $A \in \mathcal{C}$. According to [7], μ is o-continuous, hence (by Gavriluţ [6]) it is decreasing convergent. If μ has the Darboux property, then it has the property $(**)$, hence it also has (o). Since μ is also R-regular, then, by Theorem 4.3.5(i), μ is non-atomic, as claimed.

(ii) We apply Remark 4.2.2(ii) and Remark 4.2.6(ii) and (iii).

□

Remark 4.4.7 Let $\mathcal{C}$ be a δ-ring and $\mu : \mathcal{C} \to \mathcal{P}_f(\mathbb{R}_+)$ the monotone multimeasure generated by a measure $m : \mathcal{C} \to \mathbb{R}_+$ (see Example 4.2.3(i)). In [10] we have observed that μ is non-atomic (has the Darboux property, respectively) if and only if the same is m. On the other hand, by Klimkin and Svistula [14], we know that if m is σ-finite and non-atomic, then m has the Darboux property. Consequently, if μ is the monotone multimeasure generated by a σ-finite measure $m : \mathcal{C} \to \mathbb{R}_+$, then μ has the Darboux property if and only if it is non-atomic.

Corollary 4.4.8 *Let be $\mathcal{C} = \mathcal{B}_0$ (or $\mathcal{B}$) and $\mu : \mathcal{C} \to \mathcal{P}_{bfc}(\mathbb{R}_+)$ be the monotone multimeasure generated by a R'-regular, σ-finite measure $m : \mathcal{C} \to \mathbb{R}_+$. Then the following statements are equivalent:*

(i) μ has the Darboux property;
(ii) μ is non-atomic;
(iii) $(**)$ for every $t \in T$, there exists $A_t \in \mathcal{C}$ so that $t \in A_t$ and $\mu(A_t) = \{0\}$.

Remark 4.4.9 Since every Baire measure $m : \mathcal{B}_0 \to \mathbb{R}_+$ is R'-regular (see [2]), then the above corollary is still valid if $\mu : \mathcal{B}_0 \to \mathcal{P}_{bfc}(\mathbb{R}_+)$ is the monotone multimeasure generated by a σ-finite measure $m : \mathcal{B}_0 \to \mathbb{R}_+$.

4.5 Conclusions

In this chapter, we have pointed out different results concerning non-atomicity and the Darboux property for Baire/Borel multivalued set functions, with special consideration of fuzzy and/or regular multisubmeasures. Also, the connection between non-atomicity and the Darboux property was established. An extension theorem by preserving the properties was obtained and characterizations of non-atomicity were given.

References

1. Belley, J., Morales, P.: Régularité d'une fonction d'ensemble à valeurs dans un groupe topologyque. Ann. Sc. Math. Québec **3**, 185–197 (1979)
2. Dinculeanu, N.: Teoria Măsurii şi Funcţii Reale (in Romanian) (ed.) Did. şi Ped., Bucureşti (1974)
3. Dinculeanu, N., Kluvánek, I.: On vector measures. Proc. Lond. Math. Soc. **17**, 505–512 (1967)
4. Dinculeanu, N., Lewis, P.W.: Regularity of Baire measures. Proc. Am. Math. Soc. **26**, 92–94 (1970)
5. Dobrakov, I.: On submeasures, I. Diss. Math. **112**, 5–35 (1974)
6. Gavriluţ, A.: Properties of regularity for multisubmeasures. An. Şt. Univ. Iaşi, Tomul L, s. I a **f.2**, 373–392 (2004)
7. Gavriluţ, A.: Regularity and o-continuity for multisubmeasures. An. Şt. Univ. Iaşi, Tomul L, s. I a **f.2**, 393–406 (2004)
8. Gavriluţ, A.: Proprietăţi de Regularitate a Multifunc ţiilor de Mulţime (in Romanian). Casa de Editură Venus, Iaşi (2006)
9. Gavriluţ, A.: Types of extensions for multisubmeasures. An. Şt. Univ. Iaşi **f.1**, 65–74 (2008)
10. Gavriluţ, A., Croitoru, A.: Pseudo-atoms and Darboux property for set multifunctions. Fuzzy Sets Syst. **161**(22), 2897–2908 (2010)
11. Halmos, P.: Measure Theory. D. van Nostrand Company Inc., New York (1950)
12. Khurana, S.S.: Extension and regularity of group-valued Baire measures. Bull. Acad. Polon. Sci., Sér. Sci. Math. Astron. Phys. **22**, 891–895 (1974)
13. Khurana, S.S.: Extensions of group-valued regular Borel measures. Math. Nachr. **97**, 159–165 (1980)
14. Klimkin, V.M., Svistula, M.G.: Darboux property of a non-additive set function. Sb. Math. **192**, 969–978 (2001)
15. Morales, P.: Regularity and extension of semigroup valued Baire measures. In: Proceeding conference Oberwolfach 1979. Lecture Notes in Mathematics, vol. 794. Springer, Berlin, 317–323 (1980)
16. Olejcek, V.: Darboux property of regular measures. Mat. Cas. **24**(3), 283–288 (1974)
17. Pap, E.: On non-additive set functions. Atti. Sem. Mat. Fis. Univ. Modena **39**, 345–360 (1991)
18. Pap, E.: Null-Additive Set Functions. Kluwer Academic Publishers, Dordrecht (1995)
19. Precupanu, A.: Some properties of the $(B - M)$-regular multimeasures. An. Şt. Univ. Iaşi **34**, 93–103 (1988)
20. Rådström, H.: An embedding theorem for spaces of convex set. Proc. Am. Math. Soc. **3**, 165–169 (1952)
21. Sundaresan, K., Day, P.W.: Regularity of group-valued Baire and Borel measures. Proc. Am. Math. Soc. **36**, 609–612 (1972)
22. Suzuki, H.: Atoms of fuzzy measures and fuzzy integrals. Fuzzy Sets Syst. **41**, 329–342 (1991)
23. Wu, C., Bo, S.: Pseudo-atoms of fuzzy and non-fuzzy measures. Fuzzy Sets Syst. **158**, 1258–1272 (2007)

Chapter 5
Atoms and pseudo-atoms for set multifunctions

The non-additive case for set (multi)functions has received in the last decades a special attention because of its applications in mathematical economics (Klein and Thompson [16]), statistics, theory of games, human decision making (Liginlal and Ow [18]), prediction of osteoporotic fractures (Pham et al. [20]). Many authors (Bandyopadhyay [2], Dinculeanu [4], Klimkin and Svistula [17], Pap [19], Precupanu [21], Rao and Rao [22], Sugeno [23], Suzuki [24], Wu and Sun [26]) treated problems of non-additive Measure Theory in their works. It is well-known the importance of Darboux property, non-atomicity, regularity, extension in measure theory (see, for example, the famous book of Aumann and Shapley [1]). In Gavriluţ [10], Gavriluţ and Croitoru [11, 12], the notions of atom, non-atomicity and Darboux property are studied for set multifunctions.

In this chapter, the concepts of pseudo-atom and Darboux property are extended to the set-valued case for set multifunctions taking values in the family $\mathcal{P}_f(X)$ of all closed nonvoid subsets of a real normed space X, with respect to the Hausdorff topology. If the notions of atom and pseudo-atom coincide for a finitely additive set function, this is not true for a set multifunction. So, to define the concept of pseudo-atom in the set-valued case is quite justified. We establish some relationships between atoms and pseudo-atoms, between non-atomicity and the Darboux property and we prove that the range of a multimeasure with the Darboux property is a convex set. If non-atomicity and Darboux property are equivalent for certain set functions, this result is not generally valid in the case of set multifunctions.

5.1 Basic notions, terminology and results

Let T be an abstract nonvoid set and $\mathcal{C}$ a ring of subsets of T.

The usage of different types of the domain $\mathcal{C}$ will be adequate to the results that will be proved and also with respect to the references.

A. Gavriluţ et al., *Atomicity through Fractal Measure Theory*,
https://doi.org/10.1007/978-3-030-29593-6_5

Definition 5.1.1 A set function $\nu : \mathcal{C} \to \overline{\mathbb{R}}_+$, with $\nu(\emptyset) = 0$, is said to be:

(i) *monotone* if $\nu(A) \leq \nu(B)$, for every $A, B \in \mathcal{C}$, with $A \subseteq B$;
(ii) a *submeasure* (in the sense of Drewnowski [6]) if ν is monotone and *subadditive*, that is, $\nu(A \cup B) \leq \nu(A) + \nu(B)$, for every $A, B \in \mathcal{C}$, with $A \cap B = \emptyset$;
(iii) *order continuous* (shortly, *o-continuous*) if $\lim_{n\to\infty} \nu(A_n) = 0$, for every $(A_n)_{n\in\mathbb{N}^*} \subseteq \mathcal{C}$, with $A_n \searrow \emptyset$;
(iv) a *Dobrakov submeasure* (Dobrakov [5]) if ν is a submeasure and it is also o-continuous.

Definition 5.1.2 Let $\nu : \mathcal{C} \to \mathbb{R}_+$ be a set function.

(i) One says that ν has *the Darboux property* (DP) if for every $A \in \mathcal{C}$ and every $p \in (0, 1)$, there exists a set $B \in \mathcal{C}$ such that $B \subseteq A$ and $\nu(B) = p\nu(A)$.
(ii) A set $A \in \mathcal{C}$ is said to be an *atom* of ν if $\nu(A) > 0$ and for every $B \in \mathcal{C}$, with $B \subseteq A$, we have $\nu(B) = 0$ or $\nu(A \backslash B) = 0$.
(iii) ν is said to be *non-atomic* (NA) (or, *atomless*) if it has no atoms (that is, for every $A \in \mathcal{C}$ with $\nu(A) > 0$, there exists $B \in \mathcal{C}$, $B \subseteq A$, such that $\nu(B) > 0$ and $\nu(A \backslash B) > 0$).
(iv) A set $A \in \mathcal{C}$ is called a *pseudo-atom* of ν if $\nu(A) > 0$ and $B \in \mathcal{C}$, $B \subseteq A$ implies $\nu(B) = 0$ or $\nu(B) = \nu(A)$.
(v) One says that ν has *the Saks property* (SP) if for every $A \in \mathcal{C}$ and every $\varepsilon > 0$, there exists a $\mathcal{C}$-partition $\{B_i\}_{i=1}^n$ of A such that $\nu(B_i) < \varepsilon$, for every $i \in \{1, \ldots, n\}$.

Remark 5.1.3

(I) ν has DP if and only if for every $A \in \mathcal{C}$ with $\nu(A) > 0$ and every $0 < t < \nu(A)$, there is $B \in \mathcal{C}$, $B \subseteq A$ such that $\nu(B) = t$.
(II) If ν is finitely additive, then $A \in \mathcal{C}$ is an atom of ν if and only if A is a pseudo-atom of ν.
(III) The Lebesgue measure on the real line $\mathbb{R}$ has no atoms.
(IV) If $X = \{1, 2, \ldots, 10\}$ and $\mathcal{C} = \mathcal{P}(X)$, then $\mu : \mathcal{P}(X) \to \mathbb{R}_+$, $\mu(A) = \text{card}A$, for every $A \subseteq X$, has each singleton $\{i\}$, $i = \overline{1, 10}$ as an atom.
(V) The Darboux property is reminiscent of the intermediate value theorem for continuous functions.

Remark 5.1.4

(I) In what concerns non-atomicity and the Darboux property, we remind the relationships established in literature for set functions. Suppose $\nu : \mathcal{C} \to \mathbb{R}_+$.

   (i) If $\mathcal{C}$ is a δ-ring and ν is a measure, then $NA \Leftrightarrow DP$ (Dinculeanu [4]).
   (ii) If $\mathcal{C}$ is a σ-algebra and ν is a Dobrakov submeasure, then $NA \Leftrightarrow DP \Leftrightarrow SP$ (Klimkin and Svistula [17]).
   (iii) If $\mathcal{C}$ is an algebra and ν is bounded, finitely additive, then $DP \Rightarrow SP \Rightarrow NA$ (Rao and Rao [22]).

(iv) If $\mathcal{C}$ is a σ-algebra and ν is finitely additive, then $DP \not\Leftrightarrow SP$ (Klimkin and Svistula [17]).
(v) If $\mathcal{C}$ is a σ-algebra and ν is a submeasure, then $SP \Rightarrow DP$ (Klimkin and Svistula [17]).

(II) Concerning the range of a non-atomic set function, the following results are known:

(i) If $\mathcal{C}$ is a σ-algebra and $\nu : \mathcal{C} \to \mathbb{R}_+$ is a non-atomic measure, then the range of ν is $R(\nu) = [0, \nu(T)]$ (Dinculeanu [4]).
(ii) If $\mathcal{C}$ is a σ-algebra and $\nu : \mathcal{C} \to \mathbb{R}_+$ is finitely additive with SP, then $R(\nu) = [0, \nu(T)]$ (Klimkin and Svistula [17]).
(iii) If $\mathcal{C}$ is a ring and $\nu : \mathcal{C} \to X$ is a vector measure with DP, then $R(\nu)$ is convex (Bandyopadhyay [2]).

Definition 5.1.5 Let $\nu : \mathcal{C} \to \mathbb{R}_+$ be a set function, with $\nu(\emptyset) = 0$. ν is said to be:

(i) *increasing convergent* (or, *continuous from below*) if $\lim_{n\to\infty} \nu(A_n) = \nu(A)$, for every sequence of sets $(A_n)_{n\in\mathbb{N}^*} \subset \mathcal{C}$ with $A_n \nearrow A \in \mathcal{C}$.
(ii) *decreasing convergent* (or, *continuous from above*) if $\lim_{n\to\infty} \nu(A_n) = \nu(A)$, for every sequence of sets $(A_n)_{n\in\mathbb{N}^*} \subset \mathcal{C}$ with $A_n \searrow A \in \mathcal{C}$ (denoted by $A_n \searrow A$).
(iii) *fuzzy* (or, *S-monotone*, i.e., monotone in the sense of Sugeno [23]) if ν is monotone, increasing convergent and decreasing convergent.

Let $(X, \|\cdot\|)$ be a real normed space, with the distance d induced by its norm.
Let h be the Pompeiu-Hausdorff metric on $\mathcal{P}_{bf}(X)$ (see Chapter 1 and also Hu and Papageorgiou [14]).

Definition 5.1.6 If $\mu : \mathcal{C} \to \mathcal{P}_0(X)$ is a set multifunction, then μ is said to be:

(i) a *multisubmeasure* Gavriluţ [7–9] if it is monotone, $\mu(\emptyset) = \{0\}$ and $\mu(A \cup B) \subseteq \mu(A) + \mu(B)$, for every $A, B \in \mathcal{C}$, with $A \cap B = \emptyset$ (or, equivalently, for every $A, B \in \mathcal{C}$);
(ii) a *multimeasure* if $\mu(\emptyset) = \{0\}$ and $\mu(A \cup B) = \mu(A) + \mu(B)$, for every $A, B \in \mathcal{C}$, with $A \cap B = \emptyset$.
(iii) If μ is a multisubmeasure (multimeasure, respectively) and a set $A \in \mathcal{C}$ has $\mu(A) = \{0\}$, then we shall say that A is a *$\{0\}$-multisubmeasure* (*$\{0\}$-multimeasure*, respectively) *set*.

Remark 5.1.7

(I) If μ is $\mathcal{P}_f(X)$-valued, then in the Definition 5.1.6-(i) and (ii), it usually appears "$\dot{+}$" instead of "+", because the sum of two closed sets is not always closed.
(II) If μ is single-valued, with $\mu(\emptyset) = \{0\}$, then the monotonicity of μ implies that $\mu(A) = \{0\}$, for all $A \in \mathcal{C}$. Therefore, the monotonicity finds a reason for set multifunctions that are not single-valued.

(III) The definition of a $\mathcal{P}_0(X)$-valued (or even $\mathcal{P}_f(X)$-valued) multi(sub) measure cannot be reduced to that of single-valued case because $\mathcal{P}_0(X)$ (and also $\mathcal{P}_f(X)$) is not a linear space since $\mathcal{P}_0(X)$ is not a group with respect to the addition "+" defined by $M + N = \{x + y; x \in M, y \in N\}$, for every $M, N \in \mathcal{P}_0(X)$.

(IV) If ν_1, ν_2 are two finite measures defined on $\mathcal{C}$, so that $\nu_1 \leq \nu_2$ and ν_2 is a probability measure, then one obtains a particular multimeasure $\mu : \mathcal{C} \to \mathcal{P}_0([0, 1])$, $\mu(A) = [\nu_1(A), \nu_2(A)]$, $\forall A \in \mathcal{C}$, which is the simplest example of a probability multimeasure. We recall that a multimeasure $M : \mathcal{C} \to \mathcal{P}_0([0, 1])$ is said to be *a probability multimeasure* if $1 \in M(T)$. These probability multimeasures are used in control, robotics and decision theory (in Bayesian estimation).

(V) Suppose $T \in \mathcal{C}$ and μ is a multisubmeasure, so that $\overline{\mu}$ is countably additive and $\overline{\mu}(T) > 0$. Then we can generate a system of upper and lower probabilities (with applications in statistical inference—Dempster [3]) in the following way:

Let be $\mathcal{A} = \{E \subset X; \mu^{-1}(E), \mu^{+1}(E) \in \mathcal{C}\}$, where for every $E \subset X$,

$$\mu^{-1}(E) = \{t \in T; \mu(\{t\}) \cap E \neq \emptyset\}$$

and $\mu^{+1}(E) = \{t \in T; \mu(\{t\}) \subset E\}$. For every $E \in \mathcal{A}$, we define *the upper probability* of E to be

$$P^*(E) = \frac{\overline{\mu}(\mu^{-1}(E))}{\overline{\mu}(T)}$$

and *the lower probability* of E to be

$$P_*(E) = \frac{\overline{\mu}(\mu^{+1}(E))}{\overline{\mu}(T)}.$$

We remark that $P^*, P_* : \mathcal{A} \to [0, 1]$ and $P_*(E) \leq P^*(E)$, for every $E \in \mathcal{A}$.

One may regard $\overline{\mu}(\mu^{-1}(E))$ as the largest possible amount of probability from the measure $\overline{\mu}$ that can be transferred to outcomes $x \in E$ and $\overline{\mu}(\mu^{+1}(E))$ as the minimal amount of probability that can be transferred to outcomes $x \in E$.

In some of our following results, we shall assume μ to be $\mathcal{P}_f(X)$-valued, when we need h to be an extended metric.

For the beginning, let us give a series of examples:

Example 5.1.8

(I) Suppose X is an AL-space (i.e., a real Banach space equipped with a lattice order relation, which is compatible with the linear structure, such that the norm $\|\cdot\|$ on X is monotone, that is, $|x| \leq |y|$ implies $\|x\| \leq \|y\|$, for every $x, y \in X$, and also satisfying the supplementary condition $\|x + y\| = \|x\| + \|y\|$, for every $x, y \in X$, with $x, y \geq 0$).

For instance, $\mathbb{R}$, $L_1(\mu)$, l_1 are usual examples of AL-spaces.

Let Λ be the positive cone of X.

As usual, by $[x, y]$ we mean the interval consisting of all $z \in X$ so that $x \leq z \leq y$.

Suppose $m : \mathcal{C} \to \Lambda$ is an arbitrary set function, with $m(\emptyset) = 0$. We consider *the induced set multifunction* $\mu : \mathcal{C} \to \mathcal{P}_{bf}(X)$, defined for every $A \in \mathcal{C}$ by $\mu(A) = [0, m(A)]$.

$h(\mu(A), \{0\}) = \sup\limits_{0 \leq x \leq m(A)} \|x\| = \|m(A)\|$, for every $A, B \in \mathcal{C}$.

If, particularly, $X = \mathbb{R}$ and $m : \mathcal{C} \to \mathbb{R}_+$ is an arbitrary set function, with $m(\emptyset) = 0$, let be the induced set multifunction is $\mu : \mathcal{C} \to \mathcal{P}_{bf}(\mathbb{R})$ defined for every $A \in \mathcal{C}$ by $\mu(A) = [0, m(A)] (\subset \mathbb{R})$. We easily observe that μ is monotone (a multisubmeasure, a multimeasure, respectively), if and only if m is monotone (a submeasure, finitely additive, respectively) due to the identities $h([0, a], [0, b]) = |a - b|$ and $|[0, a]| = a$, for every $a, b \in \mathbb{R}_+$. μ is called *the multisubmeasure* (*multimeasure*, respectively) *induced by* m.

We thus remark that the set-valued framework is a very good direction study, because when we use a proper set multifunction (for example, the induced set multifunction), it allows us to apply our considerations to the framework of set functions taking values in important particular spaces, as, for instance, AL-spaces.

(II) Suppose $(\Omega, \mathcal{F})$ is a probability space.

As it is well known, in Dempster-Shafer mathematical theory of evidence, belief functions *Belief* (Bel) and *Plausibility* (Pl) are defined by a probability distribution $m : \mathcal{P}(\Omega) \to [0, 1]$, with $m(\emptyset) = 0$ and $\sum\limits_{A \subseteq \Omega} m(A) = 1$.

For every $A \subseteq \Omega$, $Bel(A) = \sum\limits_{B \subseteq A} m(B)$ and $Pl(A) = \sum\limits_{B, B \cap A \neq \emptyset} m(B)$.

We recall that:

(i) $Bel(A) + Pl(cA) = 1$; $Bel(A) \leq Pl(A)$.

(ii) $Bel(\Omega) = 1$, $Bel(\emptyset) = 0$, $Pl(\Omega) = 1$, $Pl(\emptyset) = 0$.

(iii) $Bel(\bigcup\limits_{i=1}^{n} A_i) \geq \sum\limits_{\emptyset \neq S \subseteq \{A_1, \ldots, A_n\}} (-1)^{|S|-1} Bel(\bigcap\limits_{A_i \in S} A_i)$ (a general version of super-additivity), for every $n \in \mathbb{N}^*$ and every $\{A_1, \ldots, A_n\} \subset \Omega$.

for $n = 2$, $Bel(A_1 \cup A_2) \geq Bel(A_1) + Bel(A_2) - Bel(A_1 \cap A_2)$.

(iv) $Pl(\bigcup\limits_{i=1}^{n} A_i) \leq \sum\limits_{\emptyset \neq S \subseteq \{A_1, \ldots, A_n\}} (-1)^{|S|-1} Pl(\bigcap\limits_{A_i \in S} A_i)$ (a general version of subadditivity).

for $n = 2$, $Pl(A_1 \cup A_2) \leq Pl(A_1) + Pl(A_2) - Pl(A_1 \cap A_2)$.

(v) Pl is increasing convergent and Bel is decreasing convergent.

Belief and Plausibility non-additive measures identify a family of probability distribution for which they are lower and upper probability measures: for every $A \subseteq \Omega$, $P(A) \in [Bel(A), Pl(A)]$.

The *Belief Interval of* A is the range defined by the minimum and maximum values which could be assigned to A: $[Bel(A), Pl(A)]$. This interval probability representation contains the precise probability of a set of interest (in the classical sense). The probability is uniquely determined if $Bel(A) = Pl(A)$. In this case, which corresponds to the classical probability, all the probabilities, $P(A)$, are uniquely determined for all subsets of Ω.

(III) (i) Let be $m_1, m_2 : \mathcal{C} \to \mathbb{R}_+$ (where $\mathcal{C}$ is a ring of subsets of an abstract space T), with $m_1(\emptyset) = 0$ and $m_2(\emptyset) = 0$. If m_1 is finitely additive and m_2 is a submeasure (finitely additive, respectively), then the set multifunction $\mu : \mathcal{C} \to \mathcal{P}_f(\mathbb{R})$, defined by

$$\mu(A) = [-m_1(A), m_2(A)], \text{ for every } A \in \mathcal{C},$$

is a multisubmeasure (a multimeasure, respectively).

We now give several examples of submeasures:

If $\mathcal{C}$ is a ring of subsets of an abstract space T and $\nu_1, \nu_2 : \mathcal{C} \to \mathbb{R}_+$ are finitely additive set functions, then the following set functions are submeasures in Drewnowski's sense [6] on $\mathcal{C}$:

$$m : \mathcal{C} \to \mathbb{R}_+, m(A) = \sqrt{\nu_1(A)}, \text{ for every } A \in \mathcal{C};$$

$$m : \mathcal{C} \to \mathbb{R}_+, m(A) = \frac{\nu_1(A)}{1 + \nu_1(A)}, \text{ for every } A \in \mathcal{C};$$

$$m : \mathcal{C} \to \mathbb{R}_+, m(A) = \max\{\nu_1(A), \nu_2(A)\}, \text{ for every } A \in \mathcal{C}.$$

(ii) Let be $\nu_1, \ldots, \nu_p : \mathcal{C} \to \mathbb{R}_+$, p finitely additive set functions, where $\mathcal{C}$ is a ring of subsets of an abstract space T.

We consider the set multifunction $\mu : \mathcal{C} \to \mathcal{P}_f(\mathbb{R})$, defined for every $A \in \mathcal{C}$ by

$$\mu(A) = \{\nu_1(A), \nu_2(A), \ldots, \nu_p(A)\}.$$

Then μ is subadditive, but it is not a multisubmeasure.

Let then be another set multifunction $\mu^{\vee} : \mathcal{C} \to \mathcal{P}_f(\mathbb{R})$, defined for every $A \in \mathcal{C}$ by:

$$\mu^{\vee}(A) = \overline{\bigcup_{\substack{B \subset A, \\ B \in \mathcal{C}}} \mu(B)}.$$

One can easily verify that $\mu^{\vee}$ is a multisubmeasure.

(iii) Let $\mathcal{C}$ be a ring of subsets of an abstract space T, $m : \mathcal{C} \to \mathbb{R}_+$ a finitely additive set function and $\mu : \mathcal{C} \to \mathcal{P}_{bf}(\mathbb{R})$ the set multifunction defined for every $A \in \mathcal{C}$ by

$$\mu(A) = \begin{cases} [-m(A), m(A)], & \text{if } m(A) \leq 1 \\ [-m(A), 1], & \text{if } m(A) > 1 \end{cases}.$$

We easily observe that μ is a multisubmeasure and $|\mu(A)| = m(A)$, for every $A \in \mathcal{C}$.

(iv) Let be $\mathcal{C} = \mathcal{P}(\mathbb{N})$ and $\mu : \mathcal{C} \to \mathcal{P}_f(\mathbb{R})$, defined for every $A \subset \mathbb{N}$ by

$$\mu(A) = \begin{cases} \{0\}, & \text{if } A \text{ is finite,} \\ [1, \infty), & \text{if } A \text{ is countable infinite.} \end{cases}$$

One can easily check that μ is not a multisubmeasure.

Definition 5.1.9 A set multifunction $\mu : \mathcal{C} \to \mathcal{P}_f(X)$, with $\mu(\emptyset) = \{0\}$, is said to be:

(i) *exhaustive* (with respect to h) if $\lim_{n\to\infty} |\mu(A_n)| = 0$, for every sequence of pairwise disjoint sets $(A_n)_{n\in\mathbb{N}^*} \subset \mathcal{C}$.
(ii) *order continuous* (shortly, *o-continuous*) (with respect to h) if $\lim_{n\to\infty} |\mu(A_n)| = 0$, for every $(A_n)_{n\in\mathbb{N}^*} \subset \mathcal{C}$, with $A_n \searrow \emptyset$.
(iii) *increasing convergent* (or *continuous from below*) if $\mu(\bigcup_{n=1}^{\infty} A_n) = \lim_{n\to\infty} \mu(A_n)$ (with respect to h), for every sequence of sets $(A_n)_{n\in\mathbb{N}^*} \subset \mathcal{C}$ with $A_n \nearrow A \in \mathcal{C}$.
(iv) *decreasing convergent* (or, *continuous from above*) if $\mu(\bigcap_{n=1}^{\infty} A_n) = \lim_{n\to\infty} \mu(A_n)$ (with respect to h), for every sequence of sets $(A_n)_{n\in\mathbb{N}^*} \subset \mathcal{C}$ with $A_n \searrow A \in \mathcal{C}$.
(v) *fuzzy* (or *S-monotone*) if μ is monotone, increasing convergent and decreasing convergent.

Remark 5.1.10

(I) If $\mu : \mathcal{C} \to \mathcal{P}_f(\mathbb{R})$ is the set multifunction induced by a set function $\nu : \mathcal{C} \to \mathbb{R}_+$ (that is, $\mu(A) = [0, \nu(A)]$, for every $A \in \mathcal{C}$), then μ is exhaustive (o-continuous, increasing convergent, decreasing convergent, fuzzy, respectively) if and only if the same is ν.
(II) In Definition 5.1.9-(v), we extended the notion of fuzzy set function to the set-valued case. The difficulty arises here since we have to consider an order relation on $\mathcal{P}_0(X)$. In our setting, we used the usual inclusion relation on $\mathcal{P}_0(X)$ (which differs from that of Guo and Zhang [13]). This seems to be more

appropriate for problems and topics in set-valued case. Our definition is also different from that of Kawake [15] because $\mathcal{P}_f(X)$ is not a linear space and so, not a Riesz space (since $\mathcal{P}_f(X)$ is not a group with respect to the Minkowski addition).

In Kawabe [15], if the Riesz space V is the real line $\mathbb{R}$, then his definition reduces to the usual fuzzy measure. In our setting, if $\mu : \mathcal{C} \to \mathcal{P}_f(\mathbb{R})$, with $\mu(\emptyset) = \{0\}$, is single-valued and monotone, then μ reduces in fact to zero function. So, our definition does not reduce to the usual single-valued case and this new setting in the set-valued case could generate interesting problems.

Remark 5.1.11

(I) If $\mathcal{C}$ is finite, then every set multifunction is increasing convergent and decreasing convergent.

(II) Let $\mu : \mathcal{C} \to \mathcal{P}_f(X)$ be a set multifunction, with $\mu(\emptyset) = \{0\}$. By the definitions, if μ is decreasing convergent, then μ is o-continuous. If μ is an o-continuous multisubmeasure, then μ is increasing convergent. Indeed, suppose μ is o-continuous. Let be $(A_n)_{n\in\mathbb{N}^*} \subset \mathcal{C}$, with $A_n \nearrow A$, where $A = \bigcup\limits_{n=1}^{\infty} A_n \in \mathcal{C}$. Then $A\backslash A_n \searrow \emptyset$, which implies $\lim\limits_{n\to\infty} |\mu(A\backslash A_n)| = 0$. Since μ is a multisubmeasure, we have $e(\mu(A), \mu(A_n)) \leq |\mu(A\backslash A_n)|$, for every $n \in \mathbb{N}^*$. It results $\lim\limits_{n\to\infty} e(\mu(A), \mu(A_n)) = 0$ and so, by the monotonicity of μ, $\lim\limits_{n\to\infty} h(\mu(A_n), \mu(A)) = 0$, which shows that μ is increasing convergent. Analogously, we can prove that μ is decreasing convergent too. Consequently, a multisubmeasure $\mu : \mathcal{C} \to \mathcal{P}_f(X)$ is o-continuous if and only if it is fuzzy.

Remark 5.1.12 If $\mathcal{C}$ is a σ-ring and $\mu : \mathcal{C} \to \mathcal{P}_f(X)$ is an o-continuous multisubmeasure, then μ is exhaustive. Indeed, let $(A_n)_{n\in\mathbb{N}^*}$ be a sequence of mutually disjoint sets of $\mathcal{C}$ and let $B_n = \bigcup\limits_{k=n}^{\infty} A_k$, for every $n \in \mathbb{N}^*$. Because $\mathcal{C}$ is a σ-ring, we have $B_n \in \mathcal{C}$, for every $n \in \mathbb{N}^*$ and $B_n \searrow \emptyset$. Since μ is o-continuous, it results $\lim\limits_{n\to\infty} |\mu(B_n)| = 0$, which implies $\lim\limits_{n\to\infty} |\mu(A_n)| = 0$ and so, μ is exhaustive.

Example 5.1.13 Let $T = \{x, y\}$, $\mathcal{C} = \mathcal{P}(T)$ and let $\nu : \mathcal{C} \to \mathbb{R}_+$ be the submeasure defined for every $A \in \mathcal{C}$ by:

$$\nu(A) = \begin{cases} 0, & \text{if } A = \emptyset, \\ 1, & \text{if } A = \{x\} \text{ or } A = \{y\}, \\ \frac{3}{2}, & \text{if } A = T. \end{cases}$$

Then the multisubmeasure μ induced by ν, defined by $\mu(A) = [0, \nu(A)]$, for every $A \in \mathcal{C}$, is fuzzy. In fact, the multisubmeasure induced by a real-valued submeasure is fuzzy if and only if the submeasure is fuzzy.

Remark 5.1.14 Let $\mu : \mathcal{C} \to \mathcal{P}_0(X)$ be a set multifunction. Then the following statements hold:

(I) $|\mu(A)| \le \overline{\mu}(A)$, for every $A \in \mathcal{C}$. So, $\overline{\mu}(A) = 0$ implies $|\mu(A)| = 0$.
(II) $\overline{\mu}$ is a monotone set function.
(III) If μ is monotone, then $|\mu|$ is also monotone.
(IV) If μ is a multisubmeasure, then $\overline{\mu}$ is finitely additive and $|\mu|$ is a submeasure.
(V) Let $A \in \mathcal{C}$. Then:

   (i) $\mu(A) = \{0\}$ if and only if $|\mu(A)| = 0$.
   (ii) If μ is monotone, then $\mu(A) = \{0\}$ if and only if $\overline{\mu}(A) = 0$.

(VI) Let $\nu : \mathcal{C} \to \mathbb{R}_+$ be a set function and $\mu : \mathcal{C} \to \mathcal{P}_f(\mathbb{R})$ defined by $\mu(A) = [0, \nu(A)]$, for every $A \in \mathcal{C}$. Then the following statements hold:

   (i) $|\mu| = \nu$.
   (ii) In addition, if ν is finitely additive, then $\overline{\mu}(A) = \nu(A)$, for every $A \in \mathcal{C}$.

5.2 Pseudo-atoms for set multifunctions

In this section, we introduce the notion of a pseudo-atom for set multifunctions. We study the relationships between atoms and pseudo-atoms and we present some properties concerning pseudo-atoms of fuzzy multisubmeasures. Refer to Section 3.2 for the definition of an atom, respectively, a non-atomic set multifunction.

In the sequel, suppose $\mathcal{C}$ is a ring of subsets of T.

Definition 5.2.1 Let $\mu : \mathcal{C} \to \mathcal{P}_0(X)$ be a set multifunction.

A set $A \in \mathcal{C}$ is called a *pseudo-atom* of μ if $\mu(A) \supsetneq \{0\}$ and for every $B \in \mathcal{C}$, with $B \subseteq A$, we have either $\mu(B) = \{0\}$ or $\mu(B) = \mu(A)$.

Remark 5.2.2

(I) Let $\mu : \mathcal{C} \to \mathcal{P}_0(X)$ be monotone, with $\mu(\emptyset) = \{0\}$ and let $A \in \mathcal{C}$. By the definitions and Remark 5.1.14-V, one can see that the following statements are equivalent:

   (a) A is an atom of μ.
   (b) A is an atom of $|\mu|$.
   (c) A is an atom of $\overline{\mu}$.

(II) Let $\mu : \mathcal{C} \to \mathcal{P}_0(X)$ be a monotone set multifunction, with $\mu(\emptyset) = \{0\}$. If $A \in \mathcal{C}$ is a pseudo-atom of μ, then A is a pseudo-atom of $|\mu|$. The converse is not always valid (even if μ is a multisubmeasure), according to the following (i). Concerning $\overline{\mu}$, there are pseudo-atoms of μ which are not pseudo-atoms of $\overline{\mu}$ and there are pseudo-atoms of $\overline{\mu}$ which are not pseudo-atoms of μ, as we can see in the following (ii) and (iii).

(i) Let be $T = \{a, b\}$, $\mathcal{C} = \mathcal{P}(T)$ and $\mu : \mathcal{C} \to \mathcal{P}_0(\mathbb{R})$ defined for every $A \in \mathcal{C}$ by:

$$\mu(A) = \begin{cases} \{0, 1, 2\}, & A = \{a, b\} \text{ or } A = \{b\} \\ \{0, 2\}, & A = \{a\} \\ \{0\}, & A = \emptyset. \end{cases}$$

We observe that μ is a multisubmeasure and

$$|\mu(A)| = \begin{cases} 2, & A \neq \emptyset \\ 0, & A = \emptyset. \end{cases}$$

Then T is a pseudo-atom of $|\mu|$, but T is not a pseudo-atom of μ. Indeed, there exists $B = \{a\} \in \mathcal{C}$, $B \subset T$, such that $\mu(B) = \{0, 2\} \supsetneq \{0\}$ and $\mu(B) \neq \mu(T)$.

(ii) Let be $T = \{a, b, c\}, \mathcal{C} = \mathcal{P}(T)$ and $\mu : \mathcal{C} \to \mathcal{P}_0(\mathbb{R})$ defined for every $A \in \mathcal{C}$ by:

$$\mu(A) = \begin{cases} [0, 2], & A \neq \emptyset \\ \{0\}, & A = \emptyset. \end{cases}$$

Then, we have $\overline{\mu}(A) = \begin{cases} 0, & A = \emptyset \\ 2, & A = \{a\} \text{ or } A = \{b\} \text{ or } A = \{c\}, \\ 4, & A = \{a, b\} \text{ or } A = \{b, c\} \text{ or } A = \{c, a\}, \\ 6, & A = T. \end{cases}$

Then $A = \{a, b\}$ is a pseudo-atom of μ, but it is not a pseudo-atom of $\overline{\mu}$. Indeed, there is $B = \{a\} \subset A$, so that $\overline{\mu}(B) = 2 > 0$ and $\overline{\mu}(B) \neq \overline{\mu}(A) = 4$.

(iii) Let be $T = \{x, y\}$, $\mathcal{C} = \mathcal{P}(T)$ and $\mu : \mathcal{C} \to \mathcal{P}_0(\mathbb{R})$ defined for every $A \in \mathcal{C}$ by:

$$\mu(A) = \begin{cases} [0, 1], & A = \{x, y\} \\ \{0, 1\}, & A = \{y\} \\ \{0\}, & A = \emptyset \text{ or } A = \{x\}. \end{cases}$$

Then, it follows:

$$\overline{\mu}(A) = \begin{cases} 1, & A \neq \emptyset \text{ and } A \neq \{x\} \\ 0, & A = \emptyset \text{ or } A = \{x\}. \end{cases}$$

Let $A = \{x, y\}$. Then A is a pseudo-atom of $\overline{\mu}$, but A is not a pseudo-atom of μ. Indeed, there is $B = \{y\} \in \mathcal{C}$, $B \subset A$, such that $\mu(B) = \{0, 1\} \supsetneq \{0\}$ and $\mu(B) \neq \mu(A)$.

(III) Let $\nu : \mathcal{C} \to \mathbb{R}_+$ be monotone, with $\nu(\emptyset) = 0$ and let $\mu : \mathcal{C} \to \mathcal{P}_0(\mathbb{R})$ be defined by $\mu(A) = [0, \nu(A)]$, for every $A \in \mathcal{C}$. If $A \in \mathcal{C}$ is a pseudo-atom of $|\mu|$, then A is a pseudo-atom of μ too. So, in this case, A is a pseudo-atom of μ if and only if it is a pseudo-atom of $|\mu|$. Also, if $\nu : \mathcal{C} \to \mathbb{R}_+$ is finitely additive, then $A \in \mathcal{C}$ is a pseudo-atom of μ if and only if it is a pseudo-atom of $\overline{\mu}/_{\mathcal{C}}$ (here, $\overline{\mu}/_{\mathcal{C}}$ is the restriction of $\overline{\mu}$ on $\mathcal{C}$, defined by $(\overline{\mu}/_{\mathcal{C}})(A) = \overline{\mu}(A)$, for every $A \in \mathcal{C}$).

Remark 5.2.3

(I) Let $\mu : \mathcal{C} \to \mathcal{P}_0(X)$ be a multisubmeasure and let $A, B \in \mathcal{C}$, with $B \subseteq A$. Then $\mu(A \backslash B) = \{0\}$ implies $\mu(A) = \mu(B)$. Indeed, by Definition 5.1.6-(ii), it results:

$$\mu(A) = \mu((A \backslash B) \cup B) \subseteq \mu(A \backslash B) + \mu(B) = \mu(B) \subseteq \mu(A).$$

So $\mu(A) = \mu(B)$.

(II) Let $\mu : \mathcal{C} \to \mathcal{P}_0(X)$ be a multimeasure and let $A, B \in \mathcal{C}$, with $B \subseteq A$. Then $\mu(A \backslash B) = \{0\}$ implies $\mu(A) = \mu(B)$.

Indeed, from Definition 5.1.6-(iii), it follows:

$$\mu(A) = \mu((A \backslash B) \cup B) = \mu(A \backslash B) + \mu(B) = \mu(B).$$

(III) If $\mu : \mathcal{C} \to \mathcal{P}_0(X)$ is a multisubmeasure (or a multimeasure), then every atom of μ is a pseudo-atom of μ. As we shall see in Examples 5.2.4-I, the converse is not valid, that is quite right for introducing the notion of pseudo-atom in the set-valued case.

(IV) Let $\mu : \mathcal{C} \to \mathcal{P}_{bf}(X)$ be a multimeasure and let $A, B \in \mathcal{C}$, with $B \subseteq A$. Then $\mu(A) = \mu(B)$ implies $\mu(A \backslash B) = \{0\}$.

It follows that every pseudo-atom of μ is an atom of μ. Consequently, $A \in \mathcal{C}$ is an atom of a multimeasure $\mu : \mathcal{C} \to \mathcal{P}_{bf}(X)$ if and only if A is a pseudo-atom of μ.

Examples 5.2.4

(I) Let $T = \{x, y\}$, $\mathcal{C} = \mathcal{P}(T)$ and for every $A \in \mathcal{C}$, let

$$\mu(A) = \begin{cases} [0, +\infty), & \text{if } A \neq \emptyset, \\ \{0\}, & \text{if } A = \emptyset. \end{cases}$$

One can easily check that $\mu : \mathcal{C} \to \mathcal{P}_0(\mathbb{R})$ is a multisubmeasure and a multimeasure.

Let $A = \{x, y\}$. There is $B = \{x\} \subseteq A$, such that $\mu(B) \supsetneq \{0\}$ and $\mu(A \backslash B) = \mu(\{y\}) = [0, +\infty) \supsetneq \{0\}$. So, A is not an atom of μ.

Now, for every $C \in \mathcal{C}$, with $C \subseteq A$, we have $\mu(C) = \{0\}$, for $C = \emptyset$ and $\mu(C) = [0, +\infty) = \mu(A)$, for $C \neq \emptyset$, which shows that A is a pseudo-atom of μ. So, there are pseudo-atoms of a multisubmeasure (or a multimeasure), which are not atoms.

(II) Let $T = \{x, y\}$, $\mathcal{C} = \mathcal{P}(T)$ and for every $A \in \mathcal{C}$, let

$$\mu(A) = \begin{cases} [0, 2], & \text{if } A = \{x, y\} \\ [0, 1], & \text{if } A = \{y\} \\ \{0\}, & \text{if } A = \emptyset \text{ or } A = \{x\}. \end{cases}$$

Evidently, μ is not a multisubmeasure.

Let $A = \{x, y\}$. There exists $B = \{y\} \subseteq A$, such that $\mu(B) = [0, 1] \supsetneq \{0\}$ and $\mu(B) \neq \mu(A)$. So A is not a pseudo-atom of μ.

Now, for every $C \in \mathcal{C}$, with $C \subseteq A$, we have:

(i) if $C = \emptyset$, then $\mu(C) = \{0\}$;
(ii) if $C = \{x\}$, then $\mu(C) = \{0\}$;
(iii) if $C = \{y\}$, then $\mu(C) = [0, 1] \supsetneq \{0\}$ and $\mu(A \backslash C) = \mu(\{x\}) = \{0\}$;
(iv) if $C = \{x, y\}$, then $\mu(A \backslash C) = \mu(\emptyset) = \{0\}$.

Consequently, A is an atom of μ.

The example shows that, if μ is not a multisubmeasure, then there are atoms which are not pseudo-atoms.

(III) Let $T = \{a, b\}$, $\mathcal{C} = \mathcal{P}(T)$ and $\nu : \mathcal{C} \to \mathbb{R}_+$, defined for every $A \in \mathcal{C}$ by

$$\nu(A) = \begin{cases} 0, & \text{if } A = \emptyset \\ 2, & \text{if } A = \{a\} \text{ or } A = \{b\} \\ 4, & \text{if } A = \{a, b\}. \end{cases}$$

Then $\{a\}$ and $\{b\}$ are atoms for the multisubmeasure μ induced by the submeasure ν, defined by $\mu(A) = [0, \nu(A)]$, for every $A \in \mathcal{C}$.

(IV) Suppose T is a countable set. Let $\mathcal{C} = \{A;\ A \subseteq T,\ A$ is finite or cA is finite$\}$ and the multisubmeasure $\mu : \mathcal{C} \to \mathcal{P}_0(\mathbb{R})$, defined for every $A \in \mathcal{C}$ by

$$\mu(A) = \begin{cases} \{0\}, & A \text{ is finite} \\ \{0, 1, 3, 9, 27\}, & cA \text{ is finite.} \end{cases}$$

Then every $A \in \mathcal{C}$, such that cA is finite, is an atom of μ.

Remark 5.2.5 Let $\mu : \mathcal{C} \to \mathcal{P}_0(X)$ be a set multifunction, with $\mu(\emptyset) = \{0\}$.

(I) Suppose μ is monotone. If $A \in \mathcal{C}$ is an atom of μ and $B \in \mathcal{C}$, $B \subseteq A$ such that $\mu(B) \supsetneq \{0\}$, then B is an atom of μ and $\mu(A \backslash B) = \{0\}$.

(II) If $A \in \mathcal{C}$ is a pseudo-atom of μ and $B \in \mathcal{C}$, $B \subseteq A$ such that $\mu(B) \supsetneq \{0\}$, then B is a pseudo-atom of μ and $\mu(B) = \mu(A)$.

(III) If $A, B \in \mathcal{C}$ are pseudo-atoms of μ and $\mu(A \cap B) \supsetneq \{0\}$, then $A \cap B$ is a pseudo-atom of μ and $\mu(A \cap B) = \mu(A) = \mu(B)$.

Remark 5.2.6 Let $\mu : \mathcal{C} \to \mathcal{P}_0(X)$ be a monotone set multifunction, with $\mu(\emptyset) = \{0\}$ and $A \in \mathcal{C}$ an atom of μ. Then $\overline{\mu}(A) = |\mu(A)|$. Indeed, according to Remark 5.1.14-I, we have $|\mu(A)| \le \overline{\mu}(A)$, for every $A \in \mathcal{C}$. Now, we observe that, if $A \in \mathcal{C}$ is an atom of μ, then $\overline{\mu}(A) \le |\mu(A)|$. Indeed, let $\{B_i\}_{i=1}^n$ be a $\mathcal{C}$-partition of A. Then there is at most one $i_0 \in \{1, \ldots, n\}$ such that $\mu(B_{i_0}) \supsetneq \{0\}$ and $\mu(B_i) = \{0\}$, for every $i \in \{1, \ldots, n\} \setminus \{i_0\}$. Since μ is monotone, we have $\sum_{i=1}^{n} |\mu(B_i)| \le |\mu(A)|$, which implies $\overline{\mu}(A) \le |\mu(A)|$, so $\overline{\mu}(A) = |\mu(A)|$.

Theorem 5.2.7 *Let $\mu : \mathcal{C} \to \mathcal{P}_0(X)$ be a multisubmeasure and let $A, B \in \mathcal{C}$ be pseudo-atoms of μ.*

(I) If $\mu(A \cap B) = \{0\}$, then $A \setminus B$ and $B \setminus A$ are pseudo-atoms of μ and $\mu(A \setminus B) = \mu(A)$, $\mu(B \setminus A) = \mu(B)$.

(II) If $\mu(A) \neq \mu(B)$, then $\mu(A \cap B) = \{0\}$, $\mu(A \setminus B) = \mu(A)$ and $\mu(B \setminus A) = \mu(B)$.

Proof

(I) We prove that $\mu(A \setminus B) \supsetneq \{0\}$. Suppose, on the contrary, that $\mu(A \setminus B) = \{0\}$. Since μ is a multisubmeasure, we obtain $\mu(A) = \mu((A \setminus B) \cup (A \cap B)) \subseteq \mu(A \setminus B) + \mu(A \cap B) = \{0\}$, which is false because A is a pseudo-atom of μ. So, $\mu(A \setminus B) \supsetneq \{0\}$. From Remark 5.2.5-II), it follows that $A \setminus B$ is a pseudo-atom of μ and $\mu(A \setminus B) = \mu(A)$. Analogously, $B \setminus A$ is a pseudo-atom of μ and $\mu(B \setminus A) = \mu(B)$.

(II) Suppose, on the contrary, that $\mu(A \cap B) \supsetneq \{0\}$. According to Remark 5.2.5-III), it results that $A \cap B$ is a pseudo-atom of μ and $\mu(A \cap B) = \mu(A) = \mu(B)$, which is false. So, we have $\mu(A \cap B) = \{0\}$ and we may use (I).

□

Theorem 5.2.8 *Let $\mu : \mathcal{C} \to \mathcal{P}_0(X)$ be a multisubmeasure and let $A, B \in \mathcal{C}$ be pseudo-atoms of μ. Then there exist mutual disjoint sets $C_1, C_2, C_3 \in \mathcal{C}$, with $A \cup B = C_1 \cup C_2 \cup C_3$, such that, for every $i \in \{1, 2, 3\}$, either C_i is a pseudo-atom of μ, or $\mu(C_i) = \{0\}$.*

Proof Let $C_1 = A \cap B$, $C_2 = A \setminus B$, $C_3 = B \setminus A$. We have the following cases:

(i) $\mu(C_1) = \{0\}$. According to Theorem 5.2.7-I, C_2 and C_3 are pseudo-atoms of μ and $\mu(C_2) = \mu(A)$, $\mu(C_3) = \mu(B)$.

(ii) $\mu(C_1) \supsetneq \{0\}$, $\mu(C_2) \supsetneq \{0\}$, $\mu(C_3) \supsetneq \{0\}$. By Remark 5.2.5-III), C_1 is a pseudo-atom of μ and $\mu(C_1) = \mu(A) = \mu(B)$. By Remark 5.2.5-II), the set C_2 is a pseudo-atom of μ. Analogously, C_3 is a pseudo-atom of μ, too.

(iii) $\mu(C_1) \supsetneq \{0\}$, $\mu(C_2) = \{0\}$, $\mu(C_3) \supsetneq \{0\}$. From Remark 5.2.5-III), it results that C_1 is a pseudo-atom of μ and $\mu(C_1) = \mu(A) = \mu(B)$. As in (ii), we obtain that C_3 is a pseudo-atom of μ and $\mu(C_3) = \mu(B)$.

The last two cases are similar to (iii).

(iv) $\mu(C_1) \supsetneq \{0\}$, $\mu(C_2) \supsetneq \{0\}$, $\mu(C_3) = \{0\}$.
(v) $\mu(C_1) \supsetneq \{0\}$, $\mu(C_2) = \mu(C_3) = \{0\}$.

□

Corollary 5.2.9 *Let $\mu : \mathcal{C} \to \mathcal{P}_0(X)$ be a multisubmeasure and let $A, B \in \mathcal{C}$ be pseudo-atoms of μ. If $\mu(A \cap B) \supsetneq \{0\}$ and $\mu(A \backslash B) = \mu(B \backslash A) = \{0\}$, then $A \cap B$ is a pseudo-atom of μ, too and $\mu(A \Delta B) = \{0\}$.*

Remark 5.2.10 Suppose $\mu : \mathcal{C} \to \mathcal{P}_f(X)$ is a multisubmeasure.

(I) According to Theorem 5.2.8 we may suppose that two pseudo-atoms of μ are either pairwise disjoint or coincide if we do not consider a $\{0\}$-multisubmeasure set.
(II) If A_i is a pseudo-atom of μ, for every $i \in \{1, \ldots, n\}$, $n \in \mathbb{N}^*$, then, analogously to Theorem 5.2.8, we can write $\bigcup\limits_{i=1}^{n} A_i = (\bigcup\limits_{j=1}^{m} B_j) \cup E$, where $\{B_j\}_{j=1}^m$ and E are pairwise disjoint sets of $\mathcal{C}$, $\{B_j\}_{j=1}^m$ are pseudo-atoms of μ and E is a $\{0\}$-multisubmeasure set.

In the sequel, using some ideas of Wu and Wu [25], we establish the following results:

Theorem 5.2.11 *Suppose $\mathcal{C}$ is a σ-ring and $\mu : \mathcal{C} \to \mathcal{P}_f(X)$ is a fuzzy multisubmeasure. Then there exist at most countable pairwise disjoint pseudo-atoms $(A_n)_{n\in\mathbb{N}^*}$ of μ which satisfy the conditions:*

(i) $|\mu(A_n)| \geq |\mu(A_{n+1})|$, $\forall n \in \mathbb{N}^*$,
(ii) $\lim\limits_{n\to\infty} |\mu(A_n)| = 0$,
(iii) $\forall \varepsilon > 0, \exists n_0 \in \mathbb{N}^*$, *such that* $|\mu(\bigcup\limits_{k=n_0}^{\infty} A_k)| < \varepsilon$.

Proof Let be $\mathcal{U}_m = \{B \in \mathcal{C} | B$ is a pseudo-atom of μ and $\frac{1}{m} \leq |\mu(B)| < \frac{1}{m+1}\}$, for every $m \in \mathbb{N}^*$. Then $\mathcal{U}_m$ contains at most finite number of sets and by Remark 5.2.10, we may suppose that they are pairwise disjoint. Suppose, on the contrary, there are infinite pairwise disjoint sets $(B_n)_{n\in\mathbb{N}} \subset \mathcal{U}_m$. So, we have $|\mu(B_n)| \geq \frac{1}{m}$, for every $n \in \mathbb{N}$. Since μ is exhaustive (according to Remarks 5.1.12 and 5.1.11-II), it follows $\lim\limits_{n\to\infty} |\mu(B_n)| = 0$, which is false. Hence, there exist at most finite pairwise disjoint pseudo-atoms in $\mathcal{U}_m$, for every $m \in \mathbb{N}^*$ and denote all of them by $\{A_n\}_{n=1}^{\infty}$. Evidently, (i) is satisfied. Since (A_n) are pairwise disjoint and μ is exhaustive, it results (ii). We remark that $\bigcap\limits_{n=1}^{\infty} \bigcup\limits_{k=n}^{\infty} A_k = \emptyset$. Since μ is o-continuous, it follows $\lim\limits_{n\to\infty} |\mu(\bigcup\limits_{k=n}^{\infty} A_k)| = 0$, which proves (iii). □

Theorem 5.2.12 *Let $\mathcal{C}$ be a σ-ring and $\mu : \mathcal{C} \to \mathcal{P}_f(X)$ a fuzzy multisubmeasure. Then there exists a sequence (A_n) of pairwise disjoint pseudo-atoms of μ, such that for every $E \in \mathcal{C}$ and every $\varepsilon > 0$, there are a subsequence (A_{i_n}) of (A_n), $p_1, p_2 \in \mathbb{N}$ and $B \in \mathcal{C}$ which is pairwise disjoint of the sets (A_{i_n}), satisfying the conditions:*

(i) $E = (\bigcup_{k=1}^{p_1} A_{i_k}) \cup (\bigcup_{k=p_1+1}^{p_2} A_{i_k}) \cup (\bigcup_{k=p_2+1}^{\infty} A_{i_k}) \cup B,$
(ii) $|\mu(A_{i_k})| > \varepsilon$, *for every* $k \in \{1, \ldots, p_1\}$,
(iii) $|\mu(A_{i_k})| \leq \varepsilon$, *for every* $k \in \{p_1 + 1, \ldots, p_2\}$,
(iv) $|\mu(\bigcup_{k=p_2+1}^{\infty} A_{i_k})| \leq \varepsilon$,
(v) B *contains no pseudo-atoms of* μ.

Proof From Theorem 5.2.8, we may suppose that all the pseudo-atoms which are contained in E are the subsequence (A_{i_n}) of (A_n) in Theorem 5.2.11. By Theorem 5.2.11-(iii), there exists $p_2 \in \mathbb{N}$ such that (iv) results. According to Theorem 5.2.11-(i), there exists $p_1 \in \mathbb{N}$ such that (ii) and (iii) follow. Now, denoting $B = E \setminus \bigcup_{n=1}^{\infty} A_{i_n}$, (i) and (v) are obtained. □

5.3 Darboux property for set multifunctions

In this section, we extend the Darboux property to set multifunctions and prove that the range of a multimeasure having Darboux property is convex. We also present some relationships among non-atomicity, Darboux property and Saks property for μ, $|\mu|$ and $\overline{\mu}$.

In the sequel, $\mathcal{C}$ is a ring of subsets of T.

Definition 5.3.1 Let $\mu : \mathcal{C} \to \mathcal{P}_0(X)$ be a set multifunction, with $\mu(\emptyset) = \{0\}$. We say that μ has *the Darboux property* (DP) if for every $A \in \mathcal{C}$, with $\mu(A) \supsetneq \{0\}$ and every $p \in (0, 1)$, there exists a set $B \in \mathcal{C}$ such that $B \subseteq A$ and $\mu(B) = p\,\mu(A)$.

Remark 5.3.2

(I) If $\mu : \mathcal{C} \to \mathcal{P}_0(X)$ is a monotone set multifunction such that $\mu(\emptyset) = \{0\}$ and if μ has DP, then $|\mu|$ has DP.
(II) Let $\nu : \mathcal{C} \to \mathbb{R}_+$ be a submeasure and let μ be the multisubmeasure induced by ν, i.e. $\mu(A) = [0, \nu(A)]$, for every $A \in \mathcal{C}$. Then μ has the Darboux property if and only if ν has the Darboux property. So, in this case, μ has DP if and only if $|\mu|$ has DP.

Proposition 5.3.3 *If $\mu : \mathcal{C} \to \mathcal{P}_0(X)$ is a multisubmeasure, such that $|\mu| : \mathcal{C} \to \mathbb{R}_+$ has the Saks property, then μ is bounded (that is, there exists $M > 0$ so that $|\mu(A)| \leq M$, for every $A \in \mathcal{C}$).*

Proof From the Saks property, for $\varepsilon = 1$, there exists a $\mathcal{C}$-partition $(B_i)_{i=1}^N$ of T such that $|\mu(B_i)| < 1$, for every $i \in \{1, \ldots, N\}$. Then, for every $A \in \mathcal{C}$, we have:

$$|\mu(A)| \le |\mu(T)| = |\mu(\bigcup_{i=1}^N B_i)| \le \sum_{i=1}^N |\mu(B_i)| < N,$$

so μ is bounded. □

The next result is a set-valued version of the famous convexity theorem of Liapunov. We establish that the range of a multimeasure, with Darboux property, is a convex set.

Theorem 5.3.4 *If a multimeasure $\mu : \mathcal{C} \to \mathcal{P}_0(X)$ has the Darboux property, then the following properties hold:*

(I) for every $Z_1, Z_2 \in \mathcal{R}(\mu) = \{\mu(A) | A \in \mathcal{C}\}$ and every $\alpha \in (0, 1)$, we have $\alpha Z_1 + (1 - \alpha) Z_2 \in \mathcal{R}(\mu)$;
(II) $R(\mu) = \bigcup_{A \in \mathcal{C}} \mu(A)$ is convex.

Proof

(I) Let $Z_1 = \mu(A)$ and $Z_2 = \mu(B)$, where $A, B \in \mathcal{C}$. Since μ has the Darboux property, there exist $C, D \in \mathcal{C}$ such that $C \subseteq A \backslash B$, $D \subseteq B \backslash A$ and $\mu(C) = \alpha\mu(A \backslash B)$, $\mu(D) = (1 - \alpha)\mu(B \backslash A)$. Then, for $E = C \cup D \cup (A \cap B) \in \mathcal{C}$, it results $\alpha Z_1 + (1 - \alpha) Z_2 = \mu(E) \in \mathcal{R}(\mu)$.
(II) Let $x_1, x_2 \in R(\mu)$ and $\alpha \in (0, 1)$. Suppose $x_1 \in \mu(A), x_2 \in \mu(B)$, with $A, B \in \mathcal{C}$. From I) it follows there is $E \in \mathcal{C}$ such that $\alpha\mu(A) + (1 - \alpha)\mu(B) = \mu(E)$. So, $\alpha x_1 + (1 - \alpha) x_2 \in R(\mu)$.

□

Further, we investigate the relationship between non-atomicity and the Darboux property for μ, $|\mu|$ and $\overline{\mu}$. These results will be synthetized in a scheme.

Theorem 5.3.5 *Let $\mu : \mathcal{C} \to \mathcal{P}_0(X)$ be a monotone set multifunction, such that $\mu(\emptyset) = \{0\}$ and $|\mu| : \mathcal{C} \to \mathbb{R}_+$.*

(I) *The following statements are equivalent:*

(a) μ is NA.
(b) $|\mu|$ is NA.
(c) $\overline{\mu}$ is NA.

(II) If μ is a multisubmeasure with the Darboux property, then μ is non-atomic and $|\mu|$ has DP.
(III) If $\mathcal{C}$ is an algebra and if $\overline{\mu}$ is bounded, finitely additive and has the Darboux property, then μ is non-atomic.

Proof

(I) The statements follow from Remark 5.2.2-I).

(II) Suppose, on the contrary, that μ has an atom $A \in \mathcal{C}$. Let be $p \in (0, 1)$. According to the Darboux property, there is $B \in \mathcal{C}$, $B \subseteq A$ such that $\mu(B) = p\mu(A)$. Since A is an atom, it follows $\mu(B) = \{0\}$ or $\mu(A \backslash B) = \{0\}$.

If $\mu(B) = \{0\}$, then $\mu(A) = \{0\}$, which is false.

If $\mu(A \backslash B) = \{0\}$, then, since μ is a multisubmeasure, we have:

$$\mu(B) \subseteq \mu(A) \subseteq \mu(A \backslash B) + \mu(B) = \mu(B). \qquad ((2))$$

So $\mu(A) = \mu(B) = p\mu(A)$. It results $|\mu(A)| = p|\mu(A)|$ and so, $\mu(A) = \{0\}$, which is false.

Consequently, μ is non-atomic. Also, by Remark 5.3.2-I), $|\mu|$ has DP.

(III) Since $\overline{\mu}$ is bounded, finitely additive on an algebra and has the Darboux property, then, by Remark 5.1.4-I) - iii), $\overline{\mu}$ is non-atomic. But this is equivalent to the non-atomicity of μ.

□

Remark 5.3.6 If a submeasure $\nu : \mathcal{C} \to \mathbb{R}_+$ has DP, then ν is NA. Indeed, consider the multisubmeasure $\mu : \mathcal{C} \to \mathcal{P}_f(\mathbb{R})$, defined by $\mu(A) = [0, \nu(A)]$, for every $A \in \mathcal{C}$. Since ν has DP, then, according to Remark 5.3.2-II), μ also has DP. So, by Theorem 5.3.5-II), μ is NA. Consequently, by Theorem 5.3.5-I), $|\mu| = \nu$ is NA.

Corollary 5.3.7 *If $\mathcal{C}$ is a σ-algebra and μ is a multisubmeasure, such that $|\mu|$ has SP, then μ is NA.*

Proof Since $|\mu|$ is a submeasure with SP on a σ-algebra, according to Remark 5.1.4-I)(v), $|\mu|$ has DP. Now, from Remark 5.3.6 and Theorem 5.3.5-I), it results that μ is NA. □

Remark 5.3.8 The converse of Theorem 5.3.5-II) is not true. Indeed, let $\mathcal{C}$ be an algebra of subsets of T, $m : \mathcal{C} \to \mathbb{R}_+$ a bounded, finitely additive set function with DP and $\mu : \mathcal{C} \to \mathcal{P}_{bf}(\mathbb{R})$ the set multifunction defined by

$$\mu(A) = \begin{cases} [-m(A), m(A)], & \text{if } m(A) \leq 1 \\ & \\ [-m(A), 1], & \text{if } m(A) > 1. \end{cases}, \text{ for every } A \in \mathcal{C}.$$

Then μ is a multisubmeasure and $|\mu| = \overline{\mu} = m$. Let us prove that μ has not the Darboux property. Suppose, on the contrary, that for every $A \in \mathcal{C}$, with $\mu(A) \supsetneq \{0\}$, and every $p \in (0, 1)$, there exists $B \in \mathcal{C}$, $B \subseteq A$ such that $\mu(B) = p\mu(A)$. Considering $m(A) > 1$, we have $p\mu(A) = [-pm(A), p]$.

We may have two situations:

(i) If $m(B) \leq 1$, then $\mu(B) = [-m(B), m(B)] = [-pm(A), p]$. It follows $m(B) = p$ and $-p = -pm(A)$. So, $m(A) = 1$. False.

(ii) If $m(B) > 1$, then we have $\mu(B) = [-m(B), 1] = [-pm(A), p]$. So $1 = p$, which is false.

So, μ has not the Darboux property, although $|\mu|$ and $\overline{\mu}$ have it.

According to Remark 5.1.4-I) (iii), m is NA, so, by Theorem 5.3.5-I), μ is NA.

Remark 5.3.9 The results of Theorem 5.3.5 are synthetized in a below scheme.

Let $\mathcal{C}$ be a σ-algebra and $\mu : \mathcal{C} \to \mathcal{P}_f(X)$ be a multisubmeasure, with $|\mu| : \mathcal{C} \to \mathbb{R}_+$. The symbol $(*)$ indicates that the implication is true if μ is the multisubmeasure induced by an o-continuous submeasure (in this case, we observe that $|\mu|$ is a Dobrakov submeasure on a σ-algebra so, by Remark 5.1.4-I) (ii), $NA \Leftrightarrow DP \Leftrightarrow SP$ for $|\mu|$).

By Remark 5.3.8, we note that there are multisubmeasures $\mu : \mathcal{C} \to \mathcal{P}_{bf}(\mathbb{R})$, with $|\mu| : \mathcal{C} \to \mathbb{R}_+$, for which non-atomicity is not equivalent to the Darboux property. In particular cases (e.g. if μ is the multisubmeasure induced by an o-continuous submeasure defined on a σ-algebra), this equivalence holds.

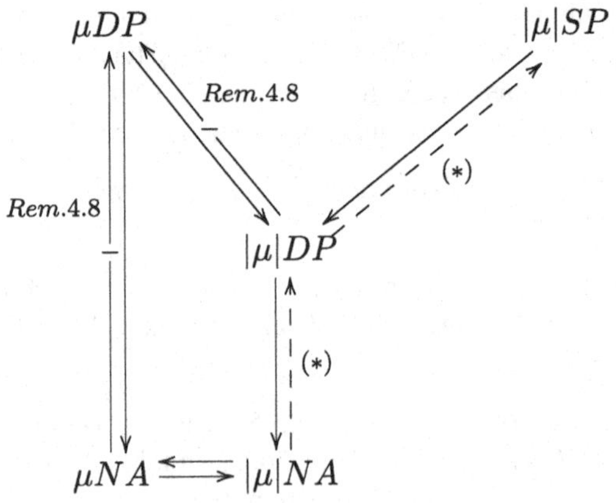

References

1. Aumann, R.J., Shapley, L.S.: Values of Non-atomic Games. Princeton University Press, Princeton (1974)
2. Bandyopadhyay, U.: On vector measures with the Darboux property. Q. J. Oxford Math. **25**, 57–61 (1974)
3. Dempster, A.P.: Upper and lower probabilities induced by a multivalued mapping. Ann. Math. Stat. **38**, 325–339 (1967)
4. Dinculeanu, N.: Vector Measures. VEB Berlin (1966)
5. Dobrakov, I.: On submeasures, I. Diss. Math. **112**, 5–35 (1974)
6. Drewnowski, L.: Topological rings of sets, continuous set functions. Integration, I, II, III, Bull. Acad. Polon. Sci. Sér. Math. Astron. Phys. 20, 269–286 (1972)

7. Gavriluţ, A.: Properties of regularity for multisubmeasures. An. Şt. Univ. Iaşi, **50**, 373–392 (2004)
8. Gavriluţ, A.: Regularity and o-continuity for multisubmeasures. An. Şt. Univ. Iaşi, **50**, 393–406 (2004)
9. Gavriluţ, A.: $\mathcal{K}$-tight multisubmeasure, $\mathcal{K} - \mathcal{D}$-regular multisubmeasure. An. Şt. Univ. "Al.I. Cuza" Iaşi, **51**, 387–404 (2005)
10. Gavriluţ, A.: Non-atomicity and the Darboux property for fuzzy and non-fuzzy Borel/Baire multivalued set functions. Fuzzy Sets Syst. **160**, 1308–1317 (2009)
11. Gavriluţ, A., Croitoru, A.: An extension by preserving non-atomicity of set multifunctions. Buletinul Institutului Politehnic din Iaşi, Secţia Matematică. Mecanică teoretică. Fizică, Tomul LIII (LVII), Fasc. **5**, 111–119 (2007)
12. Gavriluţ, A., Croitoru, A.: Non-atomicity for fuzzy and non-fuzzy multivalued set functions. Fuzzy Sets Syst. **160**, 2106–2116 (2009)
13. Guo, C., Zhang, D.: On the set-valued fuzzy measures. Inform. Sci. **160**, 13–25 (2004)
14. Hu, S., Papageorgiou, N.S.: Handbook of Multivalued Analysis, vol. I. Kluwer Academic Publishers, Dordrecht (1997)
15. Kawabe, J.: Regularity and Lusin's theorem for Riesz space-valued fuzzy measures. Fuzzy Sets Syst. **158**, 895–903 (2007)
16. Klein, E., Thompson, A.: Theory of Correspondences. Wiley, New York (1984)
17. Klimkin, V.M., Svistula, M.G.: Darboux property of a non-additive set function. Sb. Math. **192**, 969–978 (2001)
18. Liginlal D., Ow T.T.: Modelling attitude to risk in human decision process: an application of fuzzy measures. Fuzzy Sets Syst. **157**, 3040–3054 (2006)
19. Pap, E.: Null-Additive Set Functions. Kluwer Academic Publishers, Dordrecht (1995)
20. Pham, T.D., Brandl, M., Nguyen, N.D., Nguyen, T.V.: Fuzzy measure of multiple risk factors in the prediction of osteoporotic fractures. In: Proceedings of the 9th WSEAS International Conference on Fuzzy Systems [FS'08], pp. 171–177. Sofia (2008)
21. Precupanu, A.M.: On the set valued additive and subadditive set functions. An. Şt. Univ. Iaşi, **29**, 41–48 (1984)
22. Rao, K.P.S.B., Rao, M.B.: Theory of Charges. Academic Press, New York (1983)
23. Sugeno, M.: Theory of fuzzy integrals and its applications. Ph. D. thesis, Tokyo Institute of Technology, 1974
24. Suzuki, H.: Atoms of fuzzy measures and fuzzy integrals. Fuzzy Sets Syst. **41**, 329–342 (1991)
25. Wu, C., Wu C.: A note on the range of null-additive fuzzy and non-fuzzy measure. Fuzzy Sets Syst. **110**, 145–148 (2000)
26. Wu, C., Sun, B.: Pseudo-atoms of fuzzy and non-fuzzy measures. Fuzzy Sets Syst. **158**, 1258–1272 (2007)

Chapter 6
Gould integrability on atoms for set multifunctions

In this chapter we study the relationships existing between total measurability in variation and Gould type fuzzy integrability for monotone set multifunctions taking valued in the family of all closed nonempty subsets of a Banach space. We give a special interest on the behaviour of the integral on atoms and on finite unions of disjoint atoms. We establish that any continuous real-valued function defined on a compact metric space is totally-measurable in the variation of a regular finitely purely atomic multisubmeasure and it is also Gould integrable with respect to regular finitely purely atomic multisubmeasures.

In the last decades, people began to intensively study the non-additive case, and, recently, the set-valued case, due to their applications in many fields: decision theory, artificial intelligence, statistics, sociology, biology, theory of games, economic mathematics. A real interest was given to atoms, pseudo-atoms, (finitely) purely atomicity and non-atomicity for set (multi)functions. We mention here the contributions of Chiţescu [5] (on finitely purely atomic measures), Dobrakov [6], Drewnowski [7] (on submeasures), Pap [18], Jiang and Suzuki [19], Suzuki [22], Wu and Bu [23] (various problems concerning different types of fuzzy (non-additive) set functions).

In the set-valued case, many researchers were made (Abbas et al. [1], Alimohammady et al. [2], Altun [3]) and different generalizations of well-known results from the classical (fuzzy) measure theory were obtained.

In the last years, a new type of integral (called the Gould integral) was intensively studied for different types of set (multi)functions: vector valued measures (Gould [16]), multimeasures (Precupanu, Croitoru [20]), multisubmeasures (Gavriluţ [8, 9]), submeasures (Gavriluţ, Petcu [14, 15]), monotone set multifunctions (called fuzzy multimeasures) (Precupanu and Croitoru [21]). In this chapter, we give a special interest to the behaviour of the integral on atoms, for finitely purely atomicity and also on a special type of measurability, called total measurability in variation, which seems to be a useful tool in the set-valued subadditive case.

A. Gavriluţ et al., *Atomicity through Fractal Measure Theory*,
https://doi.org/10.1007/978-3-030-29593-6_6

6.1 Basic notions, terminology and results

We list the notations and several definitions and results used throughout this chapter.

Let T be an abstract nonvoid space, $\mathcal{C}$ a ring of subsets of T, X a real normed space.

If X is a Banach space, then $\mathcal{P}_{bf}(X)$ is a complete metric space [17].

h denotes the Hausdorff-Pompeiu pseudo metric on $\mathcal{P}_f(X)$ (see Chapter 1).

On $\mathcal{P}_0(X)$ we consider the Minkowski addition " $\overset{\bullet}{+}$ " [17] (see also Chapter 3).

Usually, the Minkowski addition is used in the definition of multi(sub)measures taking values in the family of nonvoid, closed subsets of X, because the classical addition of two closed sets is not, generally, a closed set, too.

Definition 6.1.1 Let $\mu : \mathcal{C} \to \mathcal{P}_f(X)$ be a set multifunction, with $\mu(\emptyset) = \{0\}$.

(I) μ is said to be:

(i) a *multisubmeasure* [8–13] if it is monotone and
$\mu(A \cup B) \subseteq \mu(A) \overset{\bullet}{+} \mu(B)$, for every $A, B \in \mathcal{C}$, with $A \cap B = \emptyset$
(or, equivalently, for every $A, B \in \mathcal{C}$);

(ii) a *multimeasure* if $\mu(A \cup B) = \mu(A) \overset{\bullet}{+} \mu(B)$, for every $A, B \in \mathcal{C}$, with $A \cap B = \emptyset$.

(iii) *null-additive* if $\mu(A \cup B) = \mu(A)$, for every $A, B \in \mathcal{C}$, with $\mu(B) = \{0\}$.

(II) μ is said to be *finitely purely atomic* if $T = \cup_{i=1}^{p} A_i$, where $A_i \in \mathcal{C}$, $i = \overline{1, p}$ are pairwise disjoint atoms of μ (evidently, here, $\mathcal{C}$ has to be an algebra).

Remark 6.1.2

(I) Suppose $m : \mathcal{C} \to \mathbb{R}_+$, with $m(\emptyset) = 0$ is an arbitrary set function.
Using the induced set multifunction $\mu : \mathcal{C} \to \mathcal{P}_{kc}(\mathbb{R}_+)$ defined for every $A \in \mathcal{C}$ by $\mu(A) = [0, m(A)]$, by the above definitions, we immediately obtain the classical corresponding notions for set functions: monotonicity, submeasure, finitely additive (null-additive, respectively) set function, atom, finitely purely atomic set function. Also, one can easily transfer by this manner the results of this chapter back to set functions.

(II) Any multisubmeasure is null-additive. The converse is not generally valid.

Suppose T is a locally compact Hausdorff space. Let $\mathcal{B}_0$ be the Baire δ-ring and $\mathcal{B}$ the Borel δ-ring. If, moreover, T is compact, then $\mathcal{B}$ is an algebra.

We recall a type of regularity that we have defined and studied in [9, 11] and [12] for different types of set multifunctions with respect to the Hausdorff topology induced by the Hausdorff pseudometric h.

Definition 6.1.3 Let $\mu : \mathcal{C} \longrightarrow \mathcal{P}_f(X)$ be a set multifunction, with $\mu(\emptyset) = \{0\}$.

(I) μ is said to be *regular* on $A \in \mathcal{C}$ (or A is said to be *regular* with respect to μ) if for every $\varepsilon > 0$, there exists a compact set $K \subset A$, $K \in \mathcal{C}$ such that $|\mu(B)| < \varepsilon$, for every $B \in \mathcal{C}$, $B \subset A \backslash K$.

(II) μ is said to be *regular* on $\mathcal{C}$ if μ is regular on every $A \in \mathcal{C}$.
The ring $\mathcal{C}$ may be considered, for instance, $\mathcal{B}_0$ or $\mathcal{B}$.
We observe that if μ is monotone, then μ is regular on $A \in \mathcal{C}$ if and only if for every $\varepsilon > 0$, there exists a compact set $K \subset A$, $K \in \mathcal{C}$ such that $|\mu(A \backslash K)| < \varepsilon$.

One can easily check the following result which generalizes the corresponding one obtained in [10] for multisubmeasures:

Theorem 6.1.4 *Let T be a locally compact, Hausdorff space and $\mu : \mathcal{B} \to \mathcal{P}_f(X)$ monotone, null-additive and regular. If $A \in \mathcal{B}$ is an atom of μ, then there exists a unique $a \in A$ so that $\mu(A \backslash \{a\}) = \{0\}$.*

Remark 6.1.5 If T is a compact, Hausdorff space, $\mu : \mathcal{B} \to \mathcal{P}_f(X)$ is a finitely purely atomic regular multisubmeasure and $A_i \in \mathcal{B}$, $i = \overline{1, p}$ are pairwise disjoint atoms of μ so that $T = \bigcup_{i=1}^{p} A_i$, then, by Theorem 6.1.4, there exist p unique points $a_1, \ldots, a_p \in T$ so that $\mu(A_i \backslash \{a_i\}) = \{0\}$, for every $i = \overline{1, p}$. Then

$$\{0\} \subseteq \mu(T \backslash \{a_1, \ldots, a_p\}) \subseteq \mu(A_1 \backslash \{a_1\}) \dot{+}$$
$$\dot{+} \ldots \dot{+} \mu(A_p \backslash \{a_p\}) = \{0\},$$

so, $\mu(T \backslash \{a_1, \ldots, a_p\}) = \{0\}$ and this implies there exist p unique points $a_1, \ldots, a_p \in T$ such that $\mu(T) = \mu(\{a_1, \ldots, a_p\})$. This means the measure of the entire space T is precisely known.

In the sequel, $\mathcal{C}$ is a ring of subsets of an abstract space T.
We consider the following extended real-valued set functions associated to a set multifunction $\mu : \mathcal{C} \longrightarrow \mathcal{P}_0(X)$, with $\mu(\emptyset) = 0$:

(i) *the variation* $\overline{\mu}$ of μ defined in Chapter 3.
μ is said to be *of finite variation* on $\mathcal{C}$ if $\overline{\mu}(A) < \infty$, for every $A \in \mathcal{C}$.
(ii) If $\mathcal{C}$ is an algebra, $\widetilde{\mu}$ defined by $\widetilde{\mu}(A) = \inf\{\overline{\mu}(B); A \subseteq B, B \in \mathcal{C}\}$, for every $A \subseteq T$.

Remark 6.1.6

(I) (i) If μ is of finite variation on $\mathcal{C}$, then $\mu : \mathcal{C} \longrightarrow \mathcal{P}_{bf}(X)$.
(ii) $\overline{\mu}$ is monotone and super-additive on $\mathcal{P}(T)$.
(II) Gavriluţ [10, 11] If μ is a multisubmeasure, then:

(i) $|\mu|$ is a submeasure in the sense of Drewnowski [7].
(ii) $\overline{\mu}$ is finitely additive on $\mathcal{C}$ and $\overline{\mu}(A) \geq |\mu(A)|$.
(iii) $\overline{\mu}(A) = |\mu(A)|$, for every atom $A \in \mathcal{C}$ of μ.

Therefore, if $\mathcal{C}$ is an algebra and $\mu : \mathcal{C} \longrightarrow \mathcal{P}_f(X)$ is a finitely purely atomic multisubmeasure, then $\overline{\mu}(T) = \sum_{i=1}^{p} |\mu(A_i|$, where $A_i \in \mathcal{C}, i = \overline{1, p}$ are pairwise disjoint atoms of μ and $T = \bigcup_{i=1}^{p} A_i$.

(III) If $\mathcal{C}$ is an algebra, then:

(i) $\widetilde{\mu}(A) = \overline{\mu}(A)$, for every $A \in \mathcal{C}$;
(ii) If μ is a multisubmeasure, then $\widetilde{\mu}$ is a submeasure on $\mathcal{P}(T)$.

6.2 Measurability on atoms

In what follows, let $\mathcal{A}$ be an algebra of subsets of the abstract space T and X be a Banach space.

Definition 6.2.1

(i) A *partition* of T is a finite family $P = \{A_i\}_{i=\overline{1,n}} \subset \mathcal{A}$ such that $A_i \cap A_j = \emptyset, i \neq j$ and $\bigcup_{i=1}^{n} A_i = T$.
(ii) Let $P = \{A_i\}_{i=\overline{1,n}}$ and $P' = \{B_j\}_{j=\overline{1,m}}$ be two partitions of T. P' is said to be *finer than* P, denoted $P \leq P'$ (or $P' \geq P$), if for every $j = \overline{1, m}$, there exists $i_j = \overline{1, n}$ so that $B_j \subseteq A_{i_j}$.
We denote by $\mathcal{P}$ the class of all partitions of T and if $A \in \mathcal{A}$ is fixed, by $\mathcal{P}_A$, the class of all partitions of A. Suppose $\mu : \mathcal{A} \longrightarrow \mathcal{P}_f(X)$, with $\mu(\emptyset) = \{0\}$ is a set multifunction.

In what follows, $f : T \to \mathbb{R}$ is a real-valued, bounded function.

Definition 6.2.2 ([12, 21])

(I) f is said to be *$\overline{\mu}$-totally-measurable (totally-measurable in the variation of μ) on $(T, \mathcal{A}, \mu)$* if for every $\varepsilon > 0$ there exists a partition $P_\varepsilon = \{A_i\}_{i=\overline{0,n}}$ of T such that:

(a) $\overline{\mu}(A_0) < \varepsilon$ and
(b) $\sup_{t,s \in A_i} |f(t) - f(s)| < \varepsilon$ for every $i = \overline{1, n}$.

(II) f is said to be *$\overline{\mu}$-totally-measurable on $B \in \mathcal{A}$* if the restriction $f|_B$ of f to B is $\overline{\mu}$-totally-measurable on $(B, \mathcal{A}_B, \mu_B)$, where $\mathcal{A}_B = \{A \cap B; A \in \mathcal{A}\}$ and $\mu_B = \mu|_{\mathcal{A}_B}$.

Remark 6.2.3 If f is $\overline{\mu}$-totally-measurable on T, then f is $\overline{\mu}$-totally-measurable on every $A \in \mathcal{A}$. Moreover,

Proposition 6.2.4 *Let us consider $A \in \mathcal{A}$ and $\{A_i\}_{i=\overline{1,p}} \subset \mathcal{A}$ so that $A = \bigcup_{i=1}^{p} A_i$.*

If $\mu : \mathcal{A} \longrightarrow \mathcal{P}_f(X)$ is a (multi) (sub)measure, then f is $\overline{\mu}$-totally-measurable on A if and only if it is $\overline{\mu}$-totally-measurable on every A_i, $i = \overline{1,p}$.

Proof According to the previous remark, the *Only if part* immediately follows. For the *If part*, one can suppose without lack of generality that $p = 2$. We denote $A_1 = M$ and $A_2 = N$.

Suppose first that $M \cap N = \emptyset$. By the $\overline{\mu}$-totally-measurability of f on M and N, there are $P_\varepsilon^M = \{M_i\}_{i=\overline{0,n}} \in \mathcal{P}_M$ and $P_\varepsilon^N = \{N_j\}_{i=\overline{0,q}} \in \mathcal{P}_N$ satisfying the conditions of totally measurability in variation.

Since μ is a (multi)(sub)measure, then $\overline{\mu}$ is additive on $\mathcal{A}$, so $P_\varepsilon^{M \cup N} = \{M_0 \cup N_0, M_1, \dots, M_n, N_1, \dots, N_q\} \in \mathcal{P}_{M\cup N}$ also satisfies the conditions of totally measurability in variation.

Consequently, f is $\overline{\mu}$-totally-measurable on $M \cup N$.

If $M \cap N \neq \emptyset$, since $M \cup N = (M \backslash N) \cup N$ and $\overline{\mu}$-totally-measurability is hereditary, the statement easily follows.

The reader is refereed to [8, 9, 12] for other properties of totally measurability in variation.

Also, it is easy to observe that f is $\overline{\mu}$-totally-measurable if and only if f is $\overline{\lambda\mu}$-totally-measurable, for every $\lambda \in \mathbb{R}$.

The reader can refer to [4] and also to Chapter 8 for different considerations concerning other types of measurability of multifunctions.

In the following, we establish that, in some conditions, continuity implies totally-measurability in variation:

Theorem 6.2.5

(I) Gavriluţ [12] *Suppose T is a compact metric space, $f : T \to \mathbb{R}$ is continuous on T and $\mu : \mathcal{B} \to \mathcal{P}_f(X)$ is regular, null-additive and monotone.*

If μ has atoms, then f is $\overline{\mu}$-totally-measurable on every atom $A_0 \in \mathcal{B}$ of μ.

(II) If, moreover, μ is a finitely purely atomic multisubmeasure, then f is $\overline{\mu}$-totally-measurable on T.

Proof

(ii) If μ is a finitely purely atomic multisubmeasure, then, according to Proposition 6.2.4 and Remark 6.1.2-II), f is $\overline{\mu}$-totally-measurable on T.

6.3 Gould integrability on atoms

In this section we point out several relationships existing between Gould integrability and total measurability in variation on atoms. We calculate the corresponding integrals on atoms and on finite unions of disjoint atoms.

Unless stated otherwise, let $\mathcal{A}$ be an algebra of subsets of an abstract space T, X be a Banach space, $\mu : \mathcal{A} \to \mathcal{P}_f(X)$ be a monotone set multifunction of finite variation, with $\mu(\emptyset) = \{0\}$ and $f : T \to \mathbb{R}$ be a bounded function.

In what follows, without any special assumptions, $\mu : \mathcal{A} \to \mathcal{P}_f(X)$ will be a monotone set multifunction, with $\mu(\emptyset) = \{0\}$, of finite variation. As before, $f : T \to \mathbb{R}$ will be a real- valued, bounded function.

$\sigma(P)$ denotes $\sum\limits_{i=1}^{n} f(t_i)\mu(A_i)$, for every partition $P = \{A_i\}_{i=\overline{1,n}}$ of T and every $t_i \in A_i, i = \overline{1,n}$.

Definition 6.3.1 ([21])

(I) f is said to be *μ-integrable on T* if the net $(\sigma(P))_{P\in(\mathcal{P},\leq)}$ is convergent in $\mathcal{P}_f(X)$, where $\mathcal{P}$, the set of all partitions of T, is ordered by the relation "$\leq$" given in Definition 6.2.1 (ii).

If $(\sigma(P))_{P\in(\mathcal{P},\leq)}$ is convergent, then its limit is called *the integral of f on T with respect to μ*, denoted by $\int_T f d\mu$.

(II) If $B \in \mathcal{A}$, f is said to be *μ-integrable on B* if the restriction $f|_B$ of f to B is μ-integrable on $(B, \mathcal{A}_B, \mu_B)$.

Remark 6.3.2 f is μ-integrable on T if and only if there is $I \in \mathcal{P}_f(X)$ such that for every $\varepsilon > 0$, there exists a partition P_ε of T, so that for every other partition of T, $P = \{A_i\}_{i=\overline{1,n}}$, with $P \geq P_\varepsilon$ and every choice of points $t_i \in A_i, i = \overline{1,n}$, we have $h(\sigma(P), I) < \varepsilon$.

Theorem 6.3.3 *Suppose $\mu : \mathcal{A} \to \mathcal{P}_f(X)$ is monotone and null-additive. If $A \in \mathcal{A}$ is an atom of μ and if f is bounded and $\overline{\mu}$-totally-measurable on A, then f is μ-integrable on A.*

Proof We observe that, if A is an atom of μ and if $\{A_i\}_{i=\overline{1,n}} \in \mathcal{P}_A$, there exists only one set, for instance, without any loss of generality, A_1, so that $\mu(A_1) \supsetneq \{0\}$ and $\mu(A_2) = \ldots = \mu(A_n) = \{0\}$.

Let $A \in \mathcal{A}$ be an atom of μ.

Since f is $\overline{\mu}$-totally-measurable on A, then for every $\varepsilon > 0$ there exists a partition $P_\varepsilon = \{A_i\}_{i=\overline{0,n}}$ of A such that:

$$\begin{cases} (i)\ \overline{\mu}(A_0) < \frac{\varepsilon}{2M} \text{ (where } M = \sup\limits_{t\in T}|f(t)| > 0 - \text{ otherwise,} \\ \qquad f \equiv 0 \text{ on } T \text{ and the proof finishes) and} \\ (ii)\ \sup\limits_{t,s\in A_i} |f(t) - f(s)| < \frac{\varepsilon}{\overline{\mu}(T)} \text{ (where } \overline{\mu}(T) > 0 - \text{ otherwise,} \\ \qquad \overline{\mu}(T) = 0, \text{ so } \mu(A) = \{0\}, \text{ for every } A \in \mathcal{A}, \text{ and,} \\ \qquad \text{consequently, there are no atoms } A \in \mathcal{A} \text{ of } \mu, \\ \qquad \text{for every } i = \overline{1,n}). \end{cases}$$

Let $\{B_j\}_{j=\overline{1,k}}, \{C_p\}_{p=\overline{1,s}} \in \mathcal{P}_A$ be two arbitrary partitions which are finer than P_ε and consider $s_j \in B_j$, $j = \overline{1,k}$, $\theta_p \in C_p$, $p = \overline{1,s}$.

We prove that

$$h({}^{\bullet}\sum_{j=1}^{k} f(s_j)\mu(B_j), {}^{\bullet}\sum_{p=1}^{s} f(\theta_p)\mu(C_p)) < \varepsilon.$$

We have two cases:

(I) $\mu(A_0) \supsetneq \{0\}$. Then $\mu(A_1) = \ldots = \mu(A_n) = \{0\}$.

Suppose, without any loss of generality that $\mu(B_1) \supsetneq \{0\}$, $\mu(C_1) \supsetneq \{0\}$ and $\mu(B_2) = \ldots = \mu(B_k) = \{0\}$, $\mu(C_2) = \ldots = \mu(C_s) = \{0\}$. Then $B_1 \subset A_0$ and $C_1 \subset A_0$. Consequently,

$$\begin{aligned} h({}^{\bullet}\sum_{j=1}^{k} f(s_j)\mu(B_j), {}^{\bullet}\sum_{p=1}^{s} f(\theta_p)\mu(C_p)) &= h(f(s_1)\mu(B_1), f(\theta_1)\mu(C_1)) \\ &\leq |f(s_1)||\mu(B_1)| + |f(\theta_1)||\mu(C_1)| \\ &\leq 2M\overline{\mu}(A_0) < \varepsilon. \end{aligned}$$

(II) $\mu(A_0) = \{0\}$. Then, without any loss of generality, $\mu(A_1) \supsetneq \{0\}$ and $\mu(A_i) = \{0\}$, for every $i = \overline{2,n}$. Suppose that $\mu(B_1) \supsetneq \{0\}$, $\mu(C_1) \supsetneq \{0\}$ and $\mu(B_2) = \ldots = \mu(B_k) = \{0\}$, $\mu(C_2) = \ldots = \mu(C_s) = \{0\}$. Then $B_1 \subset A_1$ and $C_1 \subset A_1$, and, therefore,

$$h({}^{\bullet}\sum_{j=1}^{k} f(s_j)\mu(B_j), {}^{\bullet}\sum_{p=1}^{s} f(\theta_p)\mu(C_p)) = h(f(s_1)\mu(B_1), f(\theta_1)\mu(C_1)).$$

Since A is an atom of μ and $\mu(B_1) \supsetneq \{0\}$, then $\mu(A \backslash B_1) = \{0\}$, so $\mu(C_1 \backslash B_1) = \{0\}$. By the null-additivity of μ, we get $\mu(C_1) = \mu(B_1)$. Then

$$\begin{aligned} h({}^{\bullet}\sum_{j=1}^{k} f(s_j)\mu(B_j), {}^{\bullet}\sum_{p=1}^{s} f(\theta_p)\mu(C_p)) &= h(f(s_1)\mu(B_1), f(\theta_1)\mu(C_1)) \\ &= h(f(s_1)\mu(B_1), f(\theta_1)\mu(B_1)). \end{aligned}$$

Because, generally, $h(\alpha M, \beta M) \leq |\alpha - \beta||M|$, for every $\alpha, \beta \in \mathbb{R}$ and every $M \in \mathcal{P}_f(X)$, we have

$$\begin{aligned} h({}^{\bullet}\sum_{j=1}^{k} f(s_j)\mu(B_j), {}^{\bullet}\sum_{p=1}^{s} f(\theta_p)\mu(C_p)) &\leq |\mu(B_1)||f(s_1) - f(\theta_1)| \\ &\leq \overline{\mu}(T)\frac{\varepsilon}{\overline{\mu}(T)} = \varepsilon. \end{aligned}$$

Therefore, $(\sigma(P))_{P\in\mathcal{P}_A}$ is a Cauchy net in the complete metric space $(\mathcal{P}_{bf}(X), h)$, hence f is μ-integrable on A.

By the properties of fuzzy Gould type integrability [21] and by the above theorem, we get:

Corollary 6.3.4 *If $\mu : \mathcal{A} \to \mathcal{P}_f(X)$ is finitely purely atomic, monotone and null-additive, then every bounded function $f : T \to \mathbb{R}$ which is $\overline{\mu}$-totally-measurable on atoms, is μ-integrable on T.*

If, moreover, $\mu : \mathcal{A} \to \mathcal{P}_{kc}(X)$, then f is μ-integrable on every set $A \in \mathcal{A}$.

Also, by Precupanu and Croitoru [21] (Theorems 2.11 and 2.14), we obtain:

Corollary 6.3.5 *If T is a compact metric space, $f : T \to \mathbb{R}$ is continuous on T and $\mu : \mathcal{B} \to \mathcal{P}_f(X)$ is finitely purely atomic, monotone, null-additive and regular, then f is μ-integrable on T.*

If, moreover, $\mu : \mathcal{B} \to \mathcal{P}_{kc}(X)$, then f is μ-integrable on every $A \in \mathcal{B}$.

In the end of this section, we give several examples:

Example 6.3.6

(I) If $m_1, m_2 : \mathcal{C} \to \mathbb{R}_+$, m_1 is a finitely additive set function and m_2 is a submeasure (finitely additive set function, respectively), then the set multifunction $\mu : \mathcal{C} \to \mathcal{P}_f(\mathbb{R}_+)$, defined for every $A \in \mathcal{C}$ by $\mu(A) = [-m_1(A), m_2(A)]$, is a multisubmeasure (multimeasure, respectively).

(II) If $T = \{t, s\}$ and $\mathcal{C} = \mathcal{P}(T)$, then the set multifunction $\mu : \mathcal{C} \to \mathcal{P}_f(\mathbb{R}_+)$, defined by $\mu(T) = [0, 4]$, $\mu(\emptyset) = \{0\}$ and $\mu(\{t\}) = \mu(\{s\}) = [0, 1]$ is null-additive and it is not a multisubmeasure/multimeasure.

(III) If $\mathcal{C}$ is a ring of subsets of T, $m : \mathcal{C} \to \mathbb{R}_+$ is finitely additive, then $\mu : \mathcal{C} \to \mathcal{P}_f(\mathbb{R}_+)$, defined for every $A \in \mathcal{C}$ by

$$\mu(A) = \begin{cases} [-m(A), m(A)], & \text{if } m(A) \leq 1 \\ [-m(A), 1], & \text{if } m(A) > 1 \end{cases} \text{ is a multisubmeasure.}$$

(IV) If $\mathcal{C}$ is a ring of subsets of T, $m : \mathcal{C} \to \mathbb{R}_+$ is a set function with $m(\emptyset) = \{0\}$ and $\mu : \mathcal{C} \to \mathcal{P}_{kc}(\mathbb{R}_+)$ is the induced set multifunction defined for every $A \in \mathcal{C}$ by $\mu(A) = [0, m(A)]$, then A is an atom of μ if and only if A is an atom of m.

(V) If T is a countable set and $\mathcal{C} = \{A \subset T; A \text{ is finite or } cA \text{ is finite}\}$, then for the multisubmeasure $\mu : \mathcal{C} \to \mathcal{P}_f(\mathbb{R}_+)$, defined for every $A \in \mathcal{C}$ by

$$\mu(A) = \begin{cases} \{0\}, & \text{if } A \text{ is finite} \\ \{0, 1\}, & \text{if } cA \text{ is finite,} \end{cases}$$

every $A \in \mathcal{C}$ so that cA is finite is an atom of μ.

Theorem 6.3.7 *Suppose T is a compact Hausdorff space and $\mu : \mathcal{B} \to \mathcal{P}_f(X)$ is monotone, null-additive and regular.*

If μ has atoms and if f is μ-integrable on an atom $A_0 \in \mathcal{B}$ of μ, then there exists a unique $a_0 \in A_0$ so that $\mu(A_0 \backslash \{a_0\}) = \{0\}$ and, in this case,

$$\int_{A_0} f d\mu = f(a_0)\mu(A_0).$$

Proof By the μ-integrability of f on A_0, for every $\varepsilon > 0$, there is $P_\varepsilon = \{B_j\}_{j=\overline{1,k}} \in \mathcal{P}_{A_0}$ so that for every $t_j \in B_j$, $j = \overline{1,k}$, we have

$$h(\int_{A_0} f d\mu, \sum_{j=1}^{k} f(t_j)\mu(B_j)) < \varepsilon.$$

Since $A_0 \in \mathcal{B}$ is an atom of μ, we may suppose without loss of generality that $\mu(B_1) = \mu(A_0)$ and $\mu(B_2) = \ldots = \mu(B_k) = \{0\}$. By Theorem 6.1.4, there exists a unique point $a_0 \in A_0$ so that $\mu(A_0 \backslash \{a_0\}) = \{0\}$. By the null-additivity, we also have $\mu(A_0) = \mu(\{a_0\})$.

If $a_0 \notin B_1$, then suppose, for instance, $a_0 \in B_2$. In this case, by the monotonicity of μ and since $\mu(B_2) = \{0\}$, we obtain $\mu(\{a_0\}) = \{0\}$, so $\mu(A_0) = \{0\}$, which is a contradiction. Consequently, $a_0 \in B_1$.

Let then be, particularly, $t_1 = a_0$. We get $h(\int_{A_0} f d\mu, f(a_0)\mu(A_0)) < \varepsilon$, for every $\varepsilon > 0$, so, finally, $\int_{A_0} f d\mu = f(a_0)\mu(A_0)$.

Corollary 6.3.8 (A Lebesgue Type Theorem) *Suppose T is a compact Hausdorff space and $\mu : \mathcal{B} \to \mathcal{P}_f(X)$ is monotone, null-additive and regular.*

If for every $n \in \mathbb{N}$, $f_n, f : T \to \mathbb{R}$ are μ-integrable on an atom $A_0 \in \mathcal{B}$ of μ and if $(f_n)_n$ pointwise converges to f, then

$$\lim_{n\to\infty} \int_{A_0} f_n d\mu = \int_{A_0} f d\mu \ \ (\textit{with respect to } h).$$

Proof By Theorem 6.3.7, $\exists! a_0 \in A_0$ so that $\mu(A_0 \backslash \{a_0\}) = \{0\}$ and, in this case, for every $n \in \mathbb{N}$, $\int_{A_0} f_n d\mu = f_n(a_0)\mu(A_0)$ and $\int_{A_0} f d\mu = f(a_0)\mu(A_0)$.

Since in fact $\mu : \mathcal{B} \to \mathcal{P}_{bf}(X)$, we have $\lim\limits_{n\to\infty} f_n(a_0)\mu(A_0) = f(a_0)\mu(A_0)$, so the conclusion holds.

By Theorem 6.3.7 and since the Gould type integral [21] is finitely additive and hereditary on $\mathcal{A}$ when μ is $\mathcal{P}_{kc}(X)$-valued, we obtain the following:

Corollary 6.3.9 *If T is a compact Hausdorff space, $\mu : \mathcal{B} \to \mathcal{P}_{kc}(X)$ is finitely purely atomic, monotone, null-additive (with $T = \bigcup_{i=1}^{p} A_i$, where $A_i \in \mathcal{B}$, $i = \overline{1,p}$ are pairwise disjoint atoms of μ) and regular and if $f : T \to \mathbb{R}$ is μ-integrable on*

T, then for every $i = \overline{1, p}$, there exist unique points $a_i \in A_i$ so that $\mu(A_i \setminus \{a_i\}) = \{0\}$ and, in this case,

$$\int_T f d\mu = f(a_1)\mu(A_1) + \ldots + f(a_p)\mu(A_p).$$

Proposition 6.3.10 *Suppose $\mu : \mathcal{A} \to \mathcal{P}_{kc}(X)$ is a multisubmeasure of finite variation, with $\mu(\emptyset) = \{0\}$ and $f : T \to \mathbb{R}$ is μ-integrable on T.*

Let $M : \mathcal{A} \to \mathcal{P}_{kc}(X)$ be defined for every $A \in \mathcal{A}$ by $M(A) = \int_A f d\mu$ (by Precupanu and Croitoru [21], M is a monotone multimeasure).

If f is $\overline{\mu}$-totally-measurable, then f is $\overline{M}$-totally-measurable.

Proof The statement easily follows since $|M(A)| \leq K\overline{\mu}(A)$, for every $A \in \mathcal{A}$, where $K = \sup_{t \in T} |f(t)|$. Therefore, if $\{B_j\}_{j=\overline{1,l}} \in \mathcal{P}_A$ is arbitrary, we have:

$$\sum_{j=1}^{l} |M(B_j)| \leq K \sum_{j=1}^{l} \overline{\mu}(B_j) = K\overline{\mu}(A),$$

whence $\overline{M}(A) \leq K\overline{\mu}(A)$, for every $A \in \mathcal{A}$, so, the conclusion follows.

In the sequel, we shall discuss total measurability in variation and Gould integrability for monotone set functions $m : \mathcal{A} \to \mathbb{R}_+$:

Remark 6.3.11 By the definitions, one can easily obtain the following results:

Let $m : \mathcal{A} \to \mathbb{R}_+$, with $m(\emptyset) = 0$ be an arbitrary monotone set function of finite variation and $f : T \to \mathbb{R}$ be a bounded function.

Consider the induced monotone set multifunction $\mu : \mathcal{A} \to \mathcal{P}_f(\mathbb{R})$, of finite variation, with $\mu(\emptyset) = \{0\}$, defined by $\mu(A) = [0, m(A)]$, for every $A \in \mathcal{A}$.

We observe that $\overline{\mu}(A) = \overline{m}(A)$.

The definition of $\overline{m}$-total-measurability is then the following:

f is said to be:

(i) *$\overline{m}$-totally-measurable on $(T, \mathcal{A}, \mu)$* if for every $\varepsilon > 0$ there exists a partition $P_\varepsilon = \{A_i\}_{i=\overline{0,n}}$ of T such that:

  (a) $\overline{m}(A_0) < \varepsilon$ and
  (b) $\sup_{t,s \in A_i} |f(t) - f(s)| < \varepsilon$, for every $i = \overline{1, n}$.

(ii) *$\overline{m}$-totally-measurable on $B \in \mathcal{A}$* if the restriction $f|_B$ of f to B is $\overline{m}$-totally-measurable on $(B, \mathcal{A}_B, m_B)$, where $\mathcal{A}_B = \{A \cap B; A \in \mathcal{A}\}$ and $m_B = m|_{\mathcal{A}_B}$.

We remark that f is $\overline{m}$-totally-measurable if and only if f is $\overline{\mu}$-totally-measurable.

We also observe that f is m-integrable on T if and only if there is $I \in \mathbb{R}$ such that for every $\varepsilon > 0$, there exists a partition P_ε of T, so that for every other partition of T, $P = \{A_i\}_{i=\overline{1,n}}$, with $P \geq P_\varepsilon$ and every choice of points $t_i \in A_i$, $i = \overline{1,n}$, we have $|\sigma(P) - I| = |\sum_{i=1}^{n} f(t_i)m(A_i) - I| < \varepsilon$. Here, $I = \int_T f dm$.

One can easily check that on every $A \in \mathcal{A}$, f is m-integrable if and only if it is μ-integrable and, in this case,

$$\int_A f d\mu = [0, \int_A f dm].$$

One may immediately derive by Precupanu and Croitoru [21] the properties of fuzzy Gould type integrability with respect to a monotone set function.

It would be interesting to study the relationships among this type of integrability and other fuzzy integrals.

In what follows, we investigate the relationships between total measurability in variation and Gould integrability for different types of set functions:

Theorem 6.3.12 *Suppose $m : \mathcal{A} \to \mathbb{R}_+$ is a monotone, null-additive set function of finite variation and $f : T \to \mathbb{R}$ is a bounded function. If $A \in \mathcal{A}$ is an atom of m and if f is m-integrable on A, then f is $\overline{m}$-totally-measurable on A.*

Proof Because f is m-integrable on A, then for every $\varepsilon > 0$, there exists $P_\varepsilon = \{A_i\}_{i=\overline{1,n}} \in \mathcal{P}_A$ so that for every other $\{B_j\}_{j=\overline{1,p}} \geq P_\varepsilon$ and every $t_j, s_j \in B_j$, we have

$$|\sum_{j=1}^{p} f(t_j)m(B_j) - \sum_{j=1}^{p} f(s_j)m(B_j)| < \varepsilon.$$

On the other hand, since $A \in \mathcal{A}$ is an atom of m, then we may suppose, without loss of generality that $m(A \setminus B_1) = 0$, $m(B_2) = m(B_3) = \ldots = m(B_p) = 0$.

Consequently, for every $\varepsilon > 0$ and every $t, s \in B_1$, $|f(t) - f(s)|m(A) < \varepsilon$, so $\sup\limits_{t,s \in B_1} |f(t) - f(s)| \leq \frac{\varepsilon}{m(A)}$.

Also, $\overline{m}(A \setminus B_1) = 0$ because $m(A \setminus B_1) = 0$ and this means f is $\overline{m}$-totally-measurable on A.

By Remark 6.3.11, Theorems 6.3.12 and 6.3.3, the latter applied for the induced set multifunction, we get:

Corollary 6.3.13 *Suppose $m : \mathcal{A} \to \mathbb{R}_+$ is monotone, null-additive and of finite variation. If $f : T \to \mathbb{R}$ is a bounded function, then on every atom $A \in \mathcal{A}$ of m, f is $\overline{m}$-totally-measurable if and only if f is m-integrable.*

Let us recall that for submeasures, total-measurability in variation and Gould integrability are equivalent on any subset of $\mathcal{A}$, not only on atoms (as we obtained here for monotone, null-additive set functions, which are more general than submeasures).

Remark 6.3.14 (Gavriluţ and Petcu [14]) If $m : \mathcal{A} \to \mathbb{R}_+$ is a submeasure of finite variation and if $f : T \to \mathbb{R}$ is a bounded function, then, on any $A \in \mathcal{A}$, f is $\overline{m}$-totally-measurable if and only if f is m-integrable and, moreover, $\int_A f dm = \int_A f d\overline{m}$.

References

1. Abbas, M., Imdad, M., Gopal, D.: ε-weak contractions in fuzzy metric spaces. Iran. J. Fuzzy Syst. **8**(5), 141–148 (2011)
2. Alimohammady, M., Ekici, E., Jafari, S., Roohi, M.: On fuzzy upper and lower contra-continuous multifunctions. Iran. J. Fuzzy Syst. **8**(3), 149–158 (2011)
3. Altun, I.: Some fixed point theorems for single and multivalued mappings on ordered non-archimedean fuzzy metric spaces. Iran. J. Fuzzy Syst. **7**(1), 91–96 (2010)
4. Castaing, C., Valadier, M.: Convex Analysis and Measurable Multifunctions. Lecture Notes in Mathematics, vol. 580. Springer, Berlin (1977)
5. Chiţescu, I.: Finitely purely atomic measures: coincidence and rigidity properties, Rend. Circ. Mat. Palermo (2) **50**(3), 455–476 (2001)
6. Dobrakov, I.: On submeasures, I. Diss. Math. **112**, 5–35 (1974)
7. Drewnowski, L.: Topological rings of sets, continuous set functions. Integration, I, II, III, Bull. Acad. Polon. Sci. Sér. Math. Astron. Phys. **20**, 269–276, 277–286 (1972)
8. Gavriluţ, A.: A Gould type integral with respect to a multisubmeasure. Math. Slovaca **58**(1), 1–20 (2008)
9. Gavriluţ, A.: On some properties of the Gould type integral with respect to a multisubmeasure. An. Şt. Univ. Iaşi, 52(1), 177–194 (2006)
10. Gavriluţ, A.: Non-atomicity and the Darboux property for fuzzy and non-fuzzy Borel/Baire multivalued set functions. Fuzzy Sets Syst. **160**, 1308–1317 (2009)
11. Gavriluţ, A.: Regularity and autocontinuity of set multifunctions. Fuzzy Sets Syst. **160**, 681–693 (2009)
12. Gavriluţ, A.: A Lusin type theorem for regular monotone uniformly autocontinuous set multifunctions. Fuzzy Sets Syst. **161**, 2909–2918 (2010)
13. Gavriluţ, A., Croitoru, A.: Non-atomicity for fuzzy and non-fuzzy multivalued set functions. Fuzzy Sets Syst. **160**, 2106–2116 (2009)
14. Gavriluţ, A., Petcu, A.: A Gould type integral with respect to a submeasure. An. Şt. Univ. Iaşi, Tomul LIII, f.2, 351–368 (2007)
15. Gavriluţ, A., Petcu, A.: Some properties of the Gould type integral with respect to a submeasure. Bul. Inst. Pol. Iaşi, LIII (LVII), Fasc. **5**, 121–131 (2007)
16. Gavriluţ, A.C.: Fuzzy Gould integrability on atoms. Iran. J. Fuzzy Syst. **8**(3), 113–124 (2011)
17. Gould, G.G.: On integration of vector-valued measures. Proc. Lond. Math. Soc. **15**, 193–225 (1965)
18. Hu, S., Papageorgiou, N.S.: Handbook of Multivalued Analysis, vol. I. Kluwer Academic Publishers, Dordrecht (1997)
19. Pap, E.: Null-additive Set Functions. Kluwer Academic Publishers, Dordrecht (1995)
20. Jiang, Q., Suzuki, H.: Fuzzy measures on metric spaces. Fuzzy Sets Syst. **83**, 99–106 (1996)
21. Precupanu, A., Croitoru, A.: A Gould type integral with respect to a multimeasure, I. An. Şt. Univ. Iaşi **48**, 165–200 (2002)
22. Precupanu, A., Gavriluţ, A., Croitoru, A.: A fuzzy Gould type integral. Fuzzy Sets Syst. **161**, 661–680 (2010)
23. Suzuki, H.: Atoms of fuzzy measures and fuzzy integrals. Fuzzy Sets Syst. **41**, 329–342 (1991)

Chapter 7
Continuity properties and Alexandroff theorem in Vietoris topology

7.1 Introduction

In this chapter, we introduce and study different continuity properties, as increasing/decreasing convergence, exhaustivity, order continuity and regularity in Vietoris topology for monotone set multifunctions taking values in the family of subsets of a Hausdorff linear topological space.

In this framework, we generalize to the set-valued case, well- known results from classical fuzzy (i.e., monotone) Measure Theory. An Alexandroff theorem in Vietoris topology and its converse are also established.

Because of numerous applications in many problems arise from many fields such as of optimization, convex analysis, economy, etc., the study of hypertopologies (Hausdorff, Vietoris, Wijsman, Fell, Attouch-Wets) has become of a great interest in the last decades. Important results concerning Vietoris topology can be found in Beer [2], Apreutesei [1], Billara [3], Hu and Papageorgiou [11], Klein and Thompson [14], Michael [15], etc.

On the other hand, different problems from fuzzy Measure Theory (especially, continuity properties) have recently been studied intensively due to their various applications in economics, games theory, artificial intelligence and many other important fields. It is well known that regularity is an important property of continuity, which connects Measure Theory and topology, approximating general Borel sets by more tractable sets, such as compact and/or open sets.

Continuity properties, such as regularity, increasing/decreasing convergence, order continuity and exhaustivity we studied in [6–9] in Hausdorff topology for $\mathcal{P}_c(X)$-valued monotone set multifunctions, require on X a structure of a real normed space, which is too restrictive in many situations. That is why, in the present chapter, these continuity properties are considered in Vietoris topology for $\mathcal{P}_0(X)$-valued monotone set multifunctions, X being, less restrictive, a Hausdorff linear topological space. Thus, important continuity properties from monotone Measure Theory were first generalized in [6–9] for $\mathcal{P}_c(X)$-valued monotone set

A. Gavriluţ et al., *Atomicity through Fractal Measure Theory*,
https://doi.org/10.1007/978-3-030-29593-6_7

multifunctions (X being a real normed space), which in turn are now generalized for $\mathcal{P}_0(X)$-valued monotone set multifunctions (X being a Hausdorff linear topological space).

In recent literature, by theoretical and practical necessities, classical problems from Measure Theory were first transferred to submeasures and then to other set functions having important asymptotic structural properties (much less restrictive than subadditivity) such as autocontinuity, null-additivity, null-null-additivity, etc. On the other hand, it became more and more clear that problems treated in the single-valued setting need to be extended in the set-valued framework. This general setting requires the use of a hypertopology. Among all known hypertopologies, the Hausdorff topology is the easiest to handle. But its major disadvantage is that it needs a structure of a real normed space for X. In problems when X is only linear topological, the Vietoris topology is the best choice.

In this context, we remark the interest recently given by many authors to different problems concerning set-valued monotone measures. For instance, Guo and Zhang [10] studied different problems from fuzzy Measure Theory, for set-valued measures in Kuratowski convergence. Other set-valued monotone considerations were recently obtained by Precupanu and Gavriluţ [18, 19]—in Hausdorff topology, Zhang and Guo [23], Zhang and Wang [24], etc.

7.2 Preliminaries

Let (X, τ) be a Hausdorff, linear topological space with the origin 0, T an **abstract, nonvoid set** and $\mathcal{C}$ a **ring of subsets of** T.

By $\mathcal{V}(0)$ we mean the system of neighbourhoods of 0 and by $\mathcal{U}(0)$ we denote a fundamental system of neighbourhoods of 0.

We recall the following notions and notations (Chapter 1, see also Hu and Papageorgiou [11, Ch. 1], Precupanu et al. [20, Ch. 1] and [21, Ch. 8]):

$M^- = \{C \in \mathcal{P}_0(X);\ M \cap C \neq \emptyset\}$, $M^+ = \{C \in \mathcal{P}_0(X);\ C \subseteq M\}$,
$\mathcal{S}_{UV} = \{D^+;\ D \in \tau\}$ (UV-for upper Vietoris) and $\mathcal{S}_{LV} = \{D^-;\ D \in \tau\}$ (LV-for lower Vietoris).

Vietoris topology is denoted by $\hat{\tau}_V$.

$\hat{\tau}_V^+$ *represents the upper Vietoris topology* and $\hat{\tau}_V^-$ *means the lower Vietoris topology.*

Definition 7.2.1 A set multifunction $\mu : \mathcal{C} \to \mathcal{P}_0(X)$, with $\mu(\emptyset) = \{0\}$ is said to be:

(i) monotone if it is monotone with respect to the inclusion of sets, i.e., $\mu(A) \subseteq \mu(B)$, for every $A, B \in \mathcal{C}$ with $A \subseteq B$.

(ii) a multisubmeasure if it is monotone and subadditive, i.e., $\mu(A \cup B) \subseteq \mu(A) + \mu(B)$, for every (disjoint) $A, B \in \mathcal{C}$.

Unless stated otherwise, all over this chapter we assume that (X, τ) **is a Hausdorff, linear topological space** and $\mu : \mathcal{C} \to \mathcal{P}_0(X)$ is **a monotone set multifunction.**

Whenever $(X, \|\cdot\|)$ is assumed to be, particularly, a real normed space, we shall use the following notions and notations:

d is the distance induced by the norm of X,τ_d, the topology induced by d and h, the Hausdorff pseudometric on $\mathcal{P}_0(X)$. h becomes a metric on $\mathcal{P}_{bc}(X)$, the family of all nonvoid closed bounded sets of X (Hu and Papageorgiou [11, Ch. 1], Precupanu et al. [20, Ch. 1] and [21, Ch. 8]).

If X is a Banach space, then $(\mathcal{P}_c(X), h)$ is a complete pseudometric space ($\mathcal{P}_c(X)$ means in this chapter the set of all nonvoid closed subsets of X).

If $\varepsilon > 0$, $a \in X$ and $M \subset X$, we denote by $S(a, \varepsilon) = \{x \in X; d(x, a) < \varepsilon\}$, $S_\varepsilon(M) = \{x \in X; \exists a \in M, d(x, a) < \varepsilon\}$. Also, in this case, $\mathcal{U}(0) = \{S(0, r)\}_{r>0}$.

Hausdorff topology $\hat{\tau}_H$ on $\mathcal{P}_c(X)$ induced by the Hausdorff pseudometric h can also be interpreted as the supremum of the *upper Hausdorff topology* $\hat{\tau}_H^+$ and the *lower Hausdorff topology* $\hat{\tau}_H^-$ (Hu and Papageorgiou [11, Ch. 1]).

If $M \in \mathcal{P}_f(X)$, then a base of neighbourhoods for $\hat{\tau}_H^+$ is $\{U_+^H(M, \varepsilon)\}_{\varepsilon>0}$, where $U_+^H(M, \varepsilon) = \{N \in \mathcal{P}_f(X); N \subset S_\varepsilon(M)\}$.

A base of neighbourhoods for $\hat{\tau}_H^-$ is $\{U_-^H(M, \varepsilon)\}_{\varepsilon>0}$, where $U_-^H(M, \varepsilon) = \{N \in \mathcal{P}_f(X); M \subset S_\varepsilon(N)\}$.

The following example emphasizes the importance of the set-valued framework:

Example 7.2.2 Suppose X is an *AL*-space (i.e., a real Banach space equipped with a lattice order relation, which is compatible with the linear structure, such that the norm $\|\cdot\|$ on X is monotone, that is, $|x| \le |y|$ implies $\|x\| \le \|y\|$, for every $x, y \in X$, and also satisfying the supplementary condition $\|x + y\| = \|x\| + \|y\|$, for every $x, y \in X$, with $x, y \ge 0$).

For instance, $\mathbb{R}$, $L_1(\mu)$, l_1 are usual examples of *AL*-spaces.

Let Λ be the positive cone of X.

As usual, by $[x, y]$ we mean the interval consisting of all $z \in X$ so that $x \le z \le y$.

Suppose $m : \mathcal{A} \to \Lambda$ is an arbitrary set function, with $m(\emptyset) = 0$. We consider *the induced set multifunction* $\mu : \mathcal{A} \to \mathcal{P}_{bc}(X)$ defined for every $A \in \mathcal{A}$ by $\mu(A) = [0, m(A)]$.

We observe that $h(\mu(A), \{0\}) = \sup\limits_{0 \le x \le m(A)} \|x\| = \|m(A)\|$, for every $A, B \in \mathcal{A}$.

We remark that the set-valued framework is a very good direction study, because when we use a proper set multifunction (for example, the induced set multifunction), it allows us to apply our considerations to the framework of set functions taking values in important particular spaces, for instance, *AL*-spaces.

7.3 Continuity properties in Vietoris topology

In this section, we introduce and study for $\mathcal{P}_0(X)$-valued monotone set multifunctions (X being a Hausdorff linear topological space), continuity properties such as increasing/decreasing convergence, order continuity and exhaustivity in Vietoris topology. Then we compare these notions to those studied in [5–8] for $\mathcal{P}_c(X)$-valued monotone set multifunctions, X being a real normed space. We shall obtain generalizations of our previous results established in [6–8] (which in turn generalize results—see Jiang and Suzuki [12], Jiang et al. [13], Pap [16], from classical monotone set function theory).

Definition 7.3.1 μ is said to be:

(I) (i) $\hat{\tau}_V^+$-increasing convergent ($\hat{\tau}_V^+$**-decreasing convergent**, respectively) if for every increasing (respectively, decreasing) sequence of sets $(A_n)_{n\in\mathbb{N}} \subset \mathcal{C}$, with $A_n \nearrow A \in \mathcal{C}$ (respectively, $A_n \searrow A \in \mathcal{C}$), $\mu(A_n) \xrightarrow{\hat{\tau}_V^+} \mu(A)$, i.e., for every $U \in \tau$, with $\mu(A) \subset U$, there exists $n_0 \in \mathbb{N}$ so that $\mu(A_n) \subset U$, for every $n \geq n_0$.

(ii) $\hat{\tau}_V^+ - (S)$-fuzzy if it is monotone, $\hat{\tau}_V^+$-increasing convergent and $\hat{\tau}_V^+$-decreasing convergent ((S)-fuzzy means here fuzziness in the sense of Sugeno [22], that is, monotonicity, continuity from below (i.e., increasing convergence) and continuity from above (i.e., decreasing convergence)).

We remark that many authors mean by a fuzzy measure, a monotone set function.

(II) (i) $\hat{\tau}_V^-$**-increasing convergent** ($\hat{\tau}_V^-$*-decreasing convergent*, respectively) if for every increasing (respectively, decreasing) sequence of sets $(A_n)_{n\in\mathbb{N}} \subset \mathcal{C}$, with $A_n \nearrow A \in \mathcal{C}$ (respectively, $A_n \searrow A \in \mathcal{C}$), $\mu(A_n) \xrightarrow{\hat{\tau}_V^-} \mu(A)$, i.e., for every $V \in \tau$, with $\mu(A) \cap V \neq \emptyset$, there exists $n_0 \in \mathbb{N}$ so that $\mu(A_n) \cap V \neq \emptyset$, for every $n \geq n_0$.

(ii) $\hat{\tau}_V^- - (S)$*-fuzzy* if it is monotone, $\hat{\tau}_V^-$-increasing convergent and $\hat{\tau}_V^-$-decreasing convergent.

(III) (i) $\hat{\tau}_V$*-increasing convergent* ($\hat{\tau}_V$*-decreasing convergent*, respectively) if for every increasing (respectively, decreasing) sequence of sets $(A_n)_{n\in\mathbb{N}} \subset \mathcal{C}$, with $A_n \nearrow A \in \mathcal{C}$ (respectively, $A_n \searrow A \in \mathcal{C}$), $\mu(A_n) \xrightarrow{\hat{\tau}_V} \mu(A)$, i.e., for every $U, V \in \tau$, with $\mu(A) \subset U$ and $\mu(A) \cap V \neq \emptyset$, there exists $n_0 \in \mathbb{N}$ so that $\mu(A_n) \subset U$ and $\mu(A_n) \cap V \neq \emptyset$, for every $n \geq n_0$.

(ii) $\hat{\tau}_V - (S)$*-fuzzy* if it is monotone, $\hat{\tau}_V$-increasing convergent and $\hat{\tau}_V$-decreasing convergent.

(IV) (i) $\hat{\tau}_V^+$**-exhaustive** ($\hat{\tau}_V^+$**-order continuous**, respectively) if for every sequence of sets $(A_n)_{n\in\mathbb{N}} \subset \mathcal{C}$ which is pairwise disjoint (respectively, decreasing, with $A_n \searrow \emptyset$), $\mu(A_n) \xrightarrow{\hat{\tau}_V^+} \{0\}$, i.e., for every $U_0 \in \mathcal{V}(0)$, there exists $n_0 \in \mathbb{N}$ so that $\mu(A_n) \subset U_0$, for every $n \geq n_0$.

(ii) $\widehat{\tau}_V^-$*-exhaustive* ($\widehat{\tau}_V^-$*-order continuous*, respectively) if for every sequence of sets $(A_n)_{n\in\mathbb{N}} \subset \mathcal{C}$ which is pairwise disjoint (respectively, decreasing, with $A_n \searrow \emptyset$), $\mu(A_n) \overset{\widehat{\tau}_V^-}{\rightarrow} \{0\}$, i.e., for every $V_0 \in \mathcal{V}(0)$, there exists $n_0 \in \mathbb{N}$ so that $\mu(A_n) \cap V_0 \neq \emptyset$, for every $n \geq n_0$.

(iii) $\widehat{\tau}_V$*-exhaustive* ($\widehat{\tau}_V$*-order continuous*, respectively) if for every sequence of sets $(A_n)_{n\in\mathbb{N}} \subset \mathcal{C}$ which is pairwise disjoint (respectively, decreasing, with $A_n \searrow \emptyset$), $\mu(A_n) \overset{\widehat{\tau}_V}{\rightarrow} \{0\}$, i.e., for every $U_0, V_0 \in \mathcal{V}(0)$, there exists $n_0 \in \mathbb{N}$ so that $\mu(A_n) \subset U_0$ and $\mu(A_n) \cap V_0 \neq \emptyset$, for every $n \geq n_0$.

Remark 7.3.2

(I) (i) Because of its monotonicity, μ is always $\widehat{\tau}_V^+$-increasing convergent and $\widehat{\tau}_V^-$-decreasing convergent. So, μ is $\widehat{\tau}_V^+ - (S)$-fuzzy if and only if it is $\widehat{\tau}_V^+$-decreasing convergent and it is $\widehat{\tau}_V^- - (S)$-fuzzy if and only if it is $\widehat{\tau}_V^-$-increasing convergent.

(ii) Since $\widehat{\tau}_V = \widehat{\tau}_V^+ \cup \widehat{\tau}_V^-$, then, by (i), μ is $\widehat{\tau}_V$-decreasing convergent if and only if it is $\widehat{\tau}_V^+$-decreasing convergent and it is $\widehat{\tau}_V$-increasing convergent if and only if it is $\widehat{\tau}_V^-$-increasing convergent. Consequently, μ is $\widehat{\tau}_V - (S)$-fuzzy if and only if it is $\widehat{\tau}_V^-$-increasing convergent and $\widehat{\tau}_V^+$-decreasing convergent (if and only if it is $\widehat{\tau}_V^+ - (S)$-fuzzy and $\widehat{\tau}_V^- - (S)$-fuzzy).

(II) Because $\mu(\emptyset) = \{0\}$ and μ is monotone, then μ is always $\widehat{\tau}_V^-$-exhaustive and $\widehat{\tau}_V^-$-order continuous. Consequently, since $\widehat{\tau}_V = \widehat{\tau}_V^+ \cup \widehat{\tau}_V^-$, then μ is $\widehat{\tau}_V$-exhaustive if and only if it is $\widehat{\tau}_V^+$-exhaustive and it is $\widehat{\tau}_V$-order continuous if and only if it is $\widehat{\tau}_V^+$-order continuous.

Based on the statements from Remark 7.3.2, we have highlighted in Definition 7.3.1 only the notions that we shall really study.

In what follows, we establish different relationships among these continuity properties in Vietoris topology.

Proposition 7.3.3

(i) If $\mathcal{C}$ is a σ-ring and μ is $\widehat{\tau}_V^+$-order continuous, then μ is $\widehat{\tau}_V^+$-exhaustive.

(ii) If μ is $\widehat{\tau}_V^+$-decreasing convergent, then μ is $\widehat{\tau}_V^+$-order continuous.

(iii) Suppose μ is a multisubmeasure.

(a) If μ is $\widehat{\tau}_V^+$-order continuous, then μ is $\widehat{\tau}_V^-$-increasing convergent.

(b) If, moreover, X is a metrizable, linear topological space, then μ is $\widehat{\tau}_V^+$-decreasing convergent.

Proof

(i) Let $(A_n)_{n\in\mathbb{N}} \subset \mathcal{C}$ be pairwise disjoint and consider for every $n \in \mathbb{N}$, $B_n = \bigcup_{k=n}^{\infty} A_k$. Then $B_n \searrow \emptyset$ and, since μ is $\widehat{\tau}_V^+$-order continuous, for every $U_0 \in \mathcal{V}(0)$, there exists $n_0 \in \mathbb{N}$ so that $\mu(B_n) \subset U_0$, for every $n \geq n_0$.

Then $\mu(A_n) \subset \mu(\bigcup_{k=n}^{\infty} A_k) = \mu(B_n) \subset U_0$, for every $n \geq n_0$, hence μ is $\widehat{\tau}_V^+$-exhaustive.

(ii) The statement follows immediately by definitions.

(iii) (a) Consider arbitrary increasing sequence $(A_n)_{n\in\mathbb{N}} \subset \mathcal{C}$ with $A_n \nearrow A \in \mathcal{C}$ and let $V \in \tau$, with $\mu(A) \cap V \neq \emptyset$.

It is sufficient to prove that there exists $n_0 \in \mathbb{N}$ so that $\mu(A_{n_0}) \cap V \neq \emptyset$. Suppose that, on the contrary, $\mu(A_n) \cap V = \emptyset$, for every $n \in \mathbb{N}$.

Since $\mu(A) \cap V \neq \emptyset$, there exists $x \in \mu(A) \cap V$, so $V = x + V_0$, where $V_0 \in \mathcal{V}(0)$. There exists $U_0 \in \mathcal{U}(0)$ so that $U_0 = -U_0$ and $U_0 \subset V_0$.

Because μ is $\widehat{\tau}_V^+$-order continuous, there exists $n_0 \in \mathbb{N}$ so that $\mu(A \backslash A_{n_0}) \subset U_0$.

Consequently, $x \in \mu(A) \subset \mu(A \backslash A_{n_0}) + \mu(A_{n_0})$, whence $x = y + z$, with $y \in \mu(A \backslash A_{n_0})$ and $z \in \mu(A_{n_0})$. Then $z \notin V$ and $y \in U_0$, so $z = x - y \in x - U_0 = x + U_0 \subset x + V_0 = V$, which is a contradiction.

(b) Consider arbitrary decreasing sequence of sets $(A_n)_{n\in\mathbb{N}} \subset \mathcal{C}$, with $A_n \searrow A \in \mathcal{C}$ and let be $U \in \tau$, with $\mu(A) \subset U$.

Since X is metrizable, there exists a countable fundamental system of neighbourhoods $\mathcal{U}(0) = \{V_n\}_n$ so that $V_{n+1} \subset V_n$, for every $n \in \mathbb{N}$.

First, we prove that there exists $W_0 \in \mathcal{U}(0)$ so that $\mu(A) + W_0 \subset U$. Indeed, suppose that, on the contrary, for every $n \in \mathbb{N}$, $\mu(A) + V_n \subsetneq U$. Consequently, for every $n \in \mathbb{N}$, there exists $x_n = y_n + z_n \notin U$, with $y_n \in \mu(A)$ and $z_n \in V_n$. Let be arbitrary $W \in \mathcal{V}(0)$. There is $\widetilde{W} \in \mathcal{U}(0)$ so that $\widetilde{W} \subset W$. Then there exists $n_0 \in \mathbb{N}$ so that $\widetilde{W} = V_{n_0}$, so, for every $n \geq n_0$, $z_n \in \widetilde{W} = V_{n_0} \subset W$. This implies $z_n \to 0$. On the other hand, since $y_n \in \mu(A) \subset U$, then $U = y_n + W_1$, where $W_1 \in \mathcal{V}(0)$. Then $x_n \notin U = y_n + W_1$, whence $z_n = x_n - y_n \notin W_1$, which is a contradiction because $z_n \to 0$. Consequently, there exists $W_0 \in \mathcal{U}(0)$ so that $\mu(A) + W_0 \subset U$.

Now, because $A_n \backslash A \searrow \emptyset$, there exists $n_1 \in \mathbb{N}$ so that, for every $n \geq n_1$, $\mu(A_n \backslash A) \subset W_0$ and this implies $\mu(A_n) \subset \mu(A) + \mu(A_n \backslash A) \subset \mu(A) + W_0 \subset U$, that is, μ is $\widehat{\tau}_V^+$-decreasing convergent. □

Theorem 7.3.4 *The following statements are equivalent:*

(i) μ *is* $\widehat{\tau}_V^+$*-exhaustive;*

(ii) *(E) for every monotone sequence of sets* $(A_n)_{n\in\mathbb{N}} \subset \mathcal{C}$ *and every* $U_0 \in \mathcal{V}(0)$*, there exists* $n_0 \in \mathbb{N}$ *so that* $\mu(A_n \Delta A_m) \subset U_0$*, for every* $n, m \geq n_0$*;*

(iii) *(E′) for every sequence of sets* $(A_n)_{n\in\mathbb{N}} \subset \mathcal{C}$ *and every* $U_0 \in \mathcal{V}(0)$*, there exists* $n_0 \in \mathbb{N}$ *so that* $\mu(A_n \backslash \bigcup_{k=1}^{n_0} A_k) \subset U_0$*, for every* $n > n_0$*.*

Proof (i) ⇔ (ii): *The If part.* Assume, without any loss of generality, that $(A_n)_{n\in\mathbb{N}}$ is increasing and suppose, by the contrary, that (E) does not hold. Consequently, there exist $U_0 \in \mathcal{V}(0)$ and an increasing sequence $(n_k)_k \subset \mathbb{N}$ so that $\mu(A_{n_k} \Delta A_{n_{k+1}}) \not\subset U_0$, for every $k \in \mathbb{N}$.

If we denote for every $k \in \mathbb{N}$, $B_{n_k} = A_{n_{k+1}} \backslash A_{n_k}$, then $\mu(B_{n_k}) = \mu(A_{n_k} \Delta A_{n_{k+1}}) \not\subseteq U_0$, which is false, because $(B_{n_k})_k \subset \mathcal{C}$ is pairwise disjoint and μ is $\widehat{\tau}_V^+$-exhaustive.

The Only if part Let $(A_n)_{n\in\mathbb{N}} \subset \mathcal{C}$ be pairwise disjoint and consider $B_n = \bigcup_{i=1}^{n} A_i$, for every $n \in \mathbb{N}$. Since $(B_n)_{n\in\mathbb{N}}$ is increasing convergent, then by (E), for every $U_0 \in \mathcal{V}(0)$, there is $n_0 \in \mathbb{N}$ so that $\mu(A_n) = \mu(B_{n+1} \backslash B_n) \subset U_0$, for every $n \geq n_0$. Consequently, μ is $\widehat{\tau}_V^+$-exhaustive.

(i) $\Leftrightarrow$ (iii): For the *Only if part,* let $(A_n)_{n\in\mathbb{N}} \subset \mathcal{C}$ be pairwise disjoint. For every $U_0 \in \mathcal{V}(0)$, there exists $n_0 \in \mathbb{N}$ so that $\mu(A_n) = \mu\big(A_n \backslash \overset{n_0}{\underset{k=1}{\cup}} A_k\big) \subset U_0$, for every $n > n_0$, that is, μ is$\widehat{\tau}_V^+$-exhaustive.

The *If part* immediately follows taking into account that (i) is equivalent to (ii), so we have to prove that $(E) \Rightarrow (E')$. Indeed, for every $(A_n)_{n\in\mathbb{N}} \subset \mathcal{C}$ and every $U_0 \in \mathcal{V}(0)$, consider $B_n = \bigcup_{i=1}^{n} A_i$, for every $n \in \mathbb{N}$. Applying (E) for the increasing sequence $(B_n)_{n\in\mathbb{N}}$, there is $n_0 \in \mathbb{N}$ so that, for every $n > n_0$, $\mu\big(B_n \backslash \overset{n_0}{\underset{i=1}{\cup}} A_i\big) = \mu(B_n \Delta B_{n_0}) \subset U_0$, so $\mu\big(A_n \backslash \overset{n_0}{\underset{i=1}{\cup}} A_i\big) \subset \mu\big(B_n \backslash \overset{n_0}{\underset{i=1}{\cup}} A_i\big) \subset U_0$, that is, (E') holds.

□

Now, we establish the following:

Lemma 7.3.5 *If μ is $\widehat{\tau}_V^-$-increasing convergent, $U_0 \in \mathcal{V}(0)$, $(A_n)_{n\in\mathbb{N}} \subset \mathcal{C}$, with $A_n \nearrow A \in \mathcal{C}$ are arbitrary and if $\mu(A_n) \subset U_0$, for every $n \in \mathbb{N}$, then $\mu(A) \subset U_0$.*

Proof Suppose that, on the contrary, $\mu(A) \subsetneq U_0$, whence, there is $x \in \mu(A)$ so that $x \notin U_0$. Since X is Hausdorff, there is an open neighborhood V of x such that $V \cap U_0 = \emptyset$.

Since $\mu(A) \cap V \neq \emptyset$ and μ is $\widehat{\tau}_V^-$-increasing convergent, there exists $n_0 \in \mathbb{N}$ so that $\mu(A_n) \cap V \neq \emptyset$, for every $n \geq n_0$.

Because $\mu(A_n) \subset U_0$, for every $n \in \mathbb{N}$, we get $V \cap U_0 \neq \emptyset$, which is a contradiction. □

By Theorem 7.3.4 and Lemma 7.3.5, we have:

Theorem 7.3.6 *If μ is $\widehat{\tau}_V^+$-exhaustive and $\widehat{\tau}_V^-$- increasing convergent, then μ is $\widehat{\tau}_V^+$-order continuous.*

Proof Take arbitrary $(A_n)_{n\in\mathbb{N}} \subset \mathcal{C}$, with $A_n \searrow \emptyset$ and $U_0 \in \mathcal{V}(0)$. Since μ is $\widehat{\tau}_V^+$-exhaustive, by Theorem 7.3.4, there exists $n_0 \in \mathbb{N}$ so that $\mu(A_{n_0} \backslash A_n) \subset U_0$, for every $n > n_0$.

On the other hand, because $A_{n_0} \backslash A_n \nearrow A_{n_0}$ and μ is $\widehat{\tau}_V^-$-increasing convergent, by Lemma 7.3.5, $\mu(A_{n_0}) \subset U_0$. Consequently, $\mu(A_n) \subset \mu(A_{n_0}) \subset U_0$, for every $n > n_0$, so μ is $\widehat{\tau}_V^+$-order continuous. □

By Theorem 7.3.6 and Proposition 7.3.3, we get:

Corollary 7.3.7 *If $\mathcal{C}$ is a σ-ring and if $\mu$is $s\widehat{\tau}_V^-$-increasing convergent, then μ is $\widehat{\tau}_V^+$-exhaustive if and only if μ is $\widehat{\tau}_V^+$-order continuous.*

In what follows in this section, suppose X is **a real normed space** and $\mu : \mathcal{C} \to \mathcal{P}_c(X)$ is **a monotone set multifunction**.

We shall compare continuity properties in Vietoris and Hausdorff, respectively, topology. Let us remark that, because of the monotonicity of μ, exhaustivity, decreasing convergence and order continuity we studied in [6–8] in $\widehat{\tau}_H$ are in fact, equivalently, properties of continuity in $\widehat{\tau}_H^+$ and increasing convergence is a property of continuity in, equivalently, $\widehat{\tau}_H^-$.

Since $\hat{\tau}_H^- = \hat{\tau}_V^-$, $\hat{\tau}_H^+ = \hat{\tau}_V^+$ and $\hat{\tau}_H = \hat{\tau}_V$ on $\mathcal{P}_k(X)$, we shall see that although for $\mathcal{P}_k(X)$-valued monotone set multifunctions, we have the equivalence between continuity properties (as exhaustivity, increasing/decreasing convergence or order continuity) in Hausdorff topology, respectively, in Vietoris topology, we can demonstrate that several similar corresponding results hold even if μ is only $\mathcal{P}_c(X)$-valued.

Definition 7.3.8 ([8, 18, 19]) A set multifunction $\mu : \mathcal{C} \to \mathcal{P}_c(X)$ is said to be:

(i) $\widehat{\tau}_H$-increasing convergent if $\lim\limits_{n\to\infty} h(\mu(A_n), \mu(A)) = 0$, for every increasing sequence of sets $(A_n)_n \subset \mathcal{C}$, with $A_n \nearrow A \in \mathcal{C}$.
(ii) $\widehat{\tau}_H$-decreasing convergent if $\lim\limits_{n\to\infty} h(\mu(A_n), \mu(A)) = 0$, for every decreasing sequence of sets $(A_n)_n \subset \mathcal{C}$, with $A_n \searrow A \in \mathcal{C}$.
(iii) $\widehat{\tau}_H$-exhaustive if $\lim\limits_{n\to\infty} |\mu(A_n)| = 0$, for every sequence of pairwise disjoint sets $(A_n)_n \subset \mathcal{C}$.
(iv) $\widehat{\tau}_H$-order continuous if $\lim\limits_{n\to\infty} |\mu(A_n)| = 0$, for every sequence of sets $(A_n)_n \subset \mathcal{C}$, with $A_n \searrow \emptyset$.

Theorem 7.3.9

(i) μ is $\widehat{\tau}_V^+$-exhaustive if and only if μ is $\widehat{\tau}_H$-exhaustive.
(ii) μ is $\widehat{\tau}_V^+$-order continuous if and only if μ is $\widehat{\tau}_H$-order continuous.

Proof

(i) *The If part.* Consider arbitrary pairwise disjoint $(A_n)_{n\in\mathbb{N}} \subset \mathcal{C}$ and $\varepsilon > 0$. Since μ is $\widehat{\tau}_V^+$-exhaustive, there exists $n_0 \in \mathbb{N}$ so that $\mu(A_n) \subset S(0, \varepsilon)$, for every $n \geq n_0$. Then $|\mu(A_n)| < \varepsilon$, for every $n \geq n_0$, so μ is $\widehat{\tau}_H$-exhaustive.

The Only if part. Since μ is $\widehat{\tau}_H$-exhaustive, for every $\varepsilon > 0$, there exists $n_0 \in \mathbb{N}$ so that $|\mu(A_n)| < \varepsilon$, for every $n \geq n_0$.

Consequently, for every $n \geq n_0$, $e(\mu(A_n), \{0\}) = \sup\limits_{x\in\mu(A_n)} ||x|| < \varepsilon$, which implies $\mu(A_n) \subset S(0, \varepsilon)$, so μ is $\widehat{\tau}_V^+$-exhaustive.
(ii) The statement follows immediately using the same argues as for (i).

□

Theorem 7.3.10

(i) *If μ is $\widehat{\tau}_V^+$-decreasing convergent, then μ is $\widehat{\tau}_H$-decreasing convergent.*
(ii) *If μ is $\widehat{\tau}_H$-increasing convergent, then μ is $\widehat{\tau}_V^-$-increasing convergent.*
(iii) *If $\mu : \mathcal{C} \to \mathcal{P}_k(X)$, then:*

(a) *μ is $\widehat{\tau}_V^+$-decreasing convergent if and only if μ is $\widehat{\tau}_H$-decreasing convergent.*
(b) *μ is $\widehat{\tau}_V^-$-increasing convergent if and only if μ is $\widehat{\tau}_H$-increasing convergent.*

Proof

(i) Take arbitrary $\varepsilon > 0$ and decreasing convergent sequence $(A_n)_{n\in\mathbb{N}} \subset \mathcal{C}$, with $A_n \searrow A \in \mathcal{C}$. Since μ is $\widehat{\tau}_V^+$-decreasing convergent, $\mu(A) \subset U = \bigcup_{x\in\mu(A)} S(x, \frac{\varepsilon}{2})$ and $U \in \tau$, there exists $n_0 \in \mathbb{N}$ so that $\mu(A_n) \subset \bigcup_{x\in\mu(A)} S(x, \frac{\varepsilon}{2})$, for every $n \geq n_0$.

Then, for every $n \geq n_0$ and every $y \in \mu(A_n)$, there exists $x_n^y \in \mu(A)$ so that $d(y, x_n^y) < \frac{\varepsilon}{2}$. So, $d(y, \mu(A)) < \frac{\varepsilon}{2}$ and, consequently, $h(\mu(A_n), \mu(A)) = e(\mu(A_n), \mu(A)) \leq \frac{\varepsilon}{2} < \varepsilon$, for every $n \geq n_0$, which means that μ is $\widehat{\tau}_H$-decreasing convergent.

(ii) Consider arbitrary increasing convergent sequence $(A_n)_{n\in\mathbb{N}} \subset \mathcal{C}$, with $A_n \nearrow A \in \mathcal{C}$ and let be $V \in \tau$, with $\mu(A) \cap V \neq \emptyset$. There exists $y \in \mu(A) \cap V$, so, there is $r > 0$ such that $S(y, r) \subset V$. Since μ is $\widehat{\tau}_H$-increasing convergent, there exists $n_0 \in \mathbb{N}$ so that $h(\mu(A_n), \mu(A)) = e(\mu(A), \mu(A_n)) < r$, for every $n \geq n_0$. Consequently, $\mu(A_n) \cap S(y, r) \neq \emptyset$, for every $n \geq n_0$ and since $S(y, r) \subset V$, then μ is $\widehat{\tau}_V^-$-increasing convergent.

(iii) The statements follow from $\widehat{\tau}_V^+ = \widehat{\tau}_H^+$ and $\widehat{\tau}_V^- = \widehat{\tau}_H^-$ on $\mathcal{P}_k(X)$.

□

7.4 Regularity and Alexandroff theorem in Vietoris topology

In this section, we introduce and study another remarkable continuity property—regularity—in Vietoris topology for $\mathcal{P}_0(X)$-valued monotone set multifunctions, X being a Hausdorff, linear topological space. Then, we compare regularity properties for $\mathcal{P}_c(X)$-valued monotone set multifunctions (X being a real normed space) in Vietoris and Hausdorff, respectively, topologies. In this context, we establish generalizations concerning some of our previous results [6–9]. An Alexandroff type theorem in Vietoris topology and its converse are also obtained.

Suppose (X, τ) **is a Hausdorff, linear topological space,** T **is a locally compact Hausdorff space,** $\mathcal{C}$ is $\mathcal{B}_0$ (respectively, $\mathcal{B}_0'$) or $\mathcal{C}$ is $\mathcal{B}$ (respectively, $\mathcal{B}'$) (see Sect. 2.3).

If T is metrizable or if it has a countable base, then any compact set $K \subset T$ is G_δ. Consequently, in this case, $\mathcal{B}_0 = \mathcal{B}$ (Dinculeanu [4, Ch. III, p. 187]) so, we also have $\mathcal{B}_0' = \mathcal{B}'$.

$\mathcal{K}$ is the family of all compact subsets of T and $\mathcal{D}$ is the family of all open subsets of T. As we shall see, regularity is a continuity property with respect to a suitable topology on $\mathcal{P}(T)$ (Dinculeanu [4, Ch. III, p. 197]):

For every $K \in \mathcal{K}$ and every $D \in \mathcal{D}$, with $K \subset D$, we denote $\mathcal{I}(K, D) = \{A \subset T; K \subset A \subset D\}$.

The family $\{\mathcal{I}(K, D)\}_{K\in\mathcal{K}D\in\mathcal{D}}$ is a base of a topology $\tilde{\tau}$ on $\mathcal{P}(T)$. We denote by $\tilde{\tau}$, the topology induced on any subfamily $\mathcal{S} \subset \mathcal{P}(T)$ of subsets of T.

By $\tilde{\tau}_l$ (respectively, $\tilde{\tau}_r$) we denote the topology induced on $\{\{A \subset T; K \subset A\}\}_{K\in\mathcal{K}}$ (respectively, $\{\{A \subset T; A \subset D\}\}_{D\in\mathcal{D}}$) (refer to Chapter 2 and also to (Dinculeanu [4, Ch. III, p. 197–198], for details)).

Definition 7.4.1 A class $\mathcal{F} \subset \mathcal{P}(T)$ is said to be *dense* in $\mathcal{P}(T)$ with respect to the topology induced by $\tilde{\tau}$ if for every $K \in \mathcal{K}$ and every $D \in \mathcal{D}$, with $K \subset D$, there is $A \in \mathcal{C}$ such that $K \subset A \subset D$.

Since T is locally compact, one can easily verify the following statements (Dinculeanu [4, Ch. III, p. 197]):

Remark 7.4.2

(1) $\mathcal{B}_0, \mathcal{B}, \mathcal{B}'_0, \mathcal{B}'$ are all dense in $\mathcal{P}(T)$ with respect to the topology induced by $\tilde{\tau}$.
(2) (i) For every $A \in \mathcal{C}$, there always exists $D \in \mathcal{D} \cap \mathcal{C}$ so that $A \subset D$.
    (ii) If $\mathcal{C}$ is $\mathcal{B}$ or $\mathcal{B}'$, then for every $A \in \mathcal{C}$, there always exist $K \in \mathcal{K} \cap \mathcal{C}$ and $D \in \mathcal{D} \cap \mathcal{C}$ so that $K \subset A \subset D$.

In the following, we define regularity with respect to the (upper/lower) Vietoris topology:

Definition 7.4.3

(I) An arbitrary set $A \in \mathcal{C}$ is said to be (with respect to μ):

(i) R_l^+*-regular* if $\mu : (\mathcal{C}, \tilde{\tau}_l) \to (\mathcal{P}_c(X), \widehat{\tau}_V^+)$ is continuous at A (i.e., $B \xrightarrow{\tilde{\tau}_l} A \Rightarrow \mu(B) \xrightarrow{\widehat{\tau}_V^+} \mu(A)$, i.e., for every $U \in \tau$, with $\mu(A) \subset U$, there exists $K \in \mathcal{K} \cap \mathcal{C}$ so that $K \subset A$ and $\mu(B) \subset U$, for every $B \in \mathcal{C}$, with $K \subset B \subset A$).

(ii) R_r^+**-regular** if $\mu : (\mathcal{C}, \tilde{\tau}_r) \to (\mathcal{P}_c(X), \widehat{\tau}_V^+)$ is continuous at A (i.e., $B \xrightarrow{\tilde{\tau}_r} A \Rightarrow \mu(B) \xrightarrow{\widehat{\tau}_V^+} \mu(A)$, i.e., for every $U \in \tau$, with $\mu(A) \subset U$, there exists $D \in \mathcal{D} \cap \mathcal{C}$ so that $A \subset D$ and $\mu(B) \subset U$, for every $B \in \mathcal{C}$, with $A \subset B \subset D$).

We observe that in fact, R_r^+-regularity is a continuity property between two spaces, both of them endowed with the upper Vietoris topology.

(iii) R^+*-regular* if $\mu : (\mathcal{C}, \tilde{\tau}) \to (\mathcal{P}_c(X), \widehat{\tau}_V^+)$ is continuous at A (i.e., $B \xrightarrow{\tilde{\tau}} A \Rightarrow \mu(B) \xrightarrow{\widehat{\tau}_V^+} \mu(A)$, i.e., for every $U \in \tau$, with $\mu(A) \subset U$, there exist $K \in \mathcal{K} \cap \mathcal{C}$ and $D \in \mathcal{D} \cap \mathcal{C}$ such that $K \subset A \subset D$ and $\mu(B) \subset U$, for every $B \in \mathcal{C}$, with $K \subset B \subset D$).

(iv) R_r^-*-regular* if $\mu : (\mathcal{C}, \tilde{\tau}_r) \to (\mathcal{P}_c(X), \widehat{\tau}_V^-)$ is continuous at A (i.e., $B \overset{\tilde{\tau}_r}{\to} A \Rightarrow \mu(B) \overset{\widehat{\tau}_V^-}{\to} \mu(A)$, i.e., for every $V \in \tau$, with $\mu(A) \cap V \neq \emptyset$, there exists $D \in \mathcal{D} \cap \mathcal{C}$ so that $A \subset D$ and $\mu(B) \cap V \neq \emptyset$, for every $B \in \mathcal{C}$, with $A \subset B \subset D$).

(v) R_l^-**-regular** if $\mu : (\mathcal{C}, \tilde{\tau}_l) \to (\mathcal{P}_c(X), \widehat{\tau}_V^-)$ is continuous at A (i.e., $B \overset{\tilde{\tau}_l}{\to} A \Rightarrow \mu(B) \overset{\widehat{\tau}_V^-}{\to} \mu(A)$, i.e., for every $V \in \tau$ with $\mu(A) \cap V \neq \emptyset$, there exists $K \in \mathcal{K} \cap \mathcal{C}$ such that $K \subset A$ and $\mu(B) \cap V \neq \emptyset$, for every $B \in \mathcal{C}$, with $K \subset B \subset A$).

(vi) R^-*-regular* if $\mu : (\mathcal{C}, \tilde{\tau}) \to (\mathcal{P}_c(X), \widehat{\tau}_V^-)$ is continuous at A (i.e., $B \overset{\tilde{\tau}}{\to} A \Rightarrow \mu(B) \overset{\widehat{\tau}_V^-}{\to} \mu(A)$, i.e., for every $V \in \tau$, with $\mu(A) \cap V \neq \emptyset$, there exist $K \in \mathcal{K} \cap \mathcal{C}$ and $D \in \mathcal{D} \cap \mathcal{C}$ so that $K \subset A \subset D$ and for every $B \in \mathcal{C}$, with $K \subset B \subset D$, we have $\mu(B) \cap V \neq \emptyset$).

(vii) R_l^V*-regular* if $\mu : (\mathcal{C}, \tilde{\tau}_l) \to (\mathcal{P}_c(X), \widehat{\tau}_V)$ is continuous at A (i.e., $B \overset{\tilde{\tau}_l}{\to} A \Rightarrow \mu(B) \overset{\widehat{\tau}_V}{\to} \mu(A)$, i.e., for every $U, V \in \tau$, with $\mu(A) \subset U$ and $\mu(A) \cap V \neq \emptyset$, there exists $K \in \mathcal{K} \cap \mathcal{C}$, $K \subset A$ so that for every $B \in \mathcal{C}$, with $K \subset B \subset A$, we have $\mu(B) \subset U$ and $\mu(B) \cap V \neq \emptyset$).

(viii) R_r^V*-regular* if $\mu : (\mathcal{C}, \tilde{\tau}_r) \to (\mathcal{P}_c(X), \widehat{\tau}_V)$ is continuous at A (i.e., $B \overset{\tilde{\tau}_r}{\to} A \Rightarrow \mu(B) \overset{\widehat{\tau}_V}{\to} \mu(A)$, i.e., for every $U, V \in \tau$, with $\mu(A) \subset U$ and $\mu(A) \cap V \neq \emptyset$, there exists $D \in \mathcal{D} \cap \mathcal{C}$, with $A \subset D$, so that for every $B \in \mathcal{C}$, with $A \subset B \subset D$, we have $\mu(B) \subset U$ and $\mu(B) \cap V \neq \emptyset$).

(ix) R^V**-regular** if $\mu : (\mathcal{C}, \tilde{\tau}) \to (\mathcal{P}_c(X), \widehat{\tau}_V)$ is continuous at A (i.e., $B \overset{\tilde{\tau}}{\to} A \Rightarrow \mu(B) \overset{\widehat{\tau}_V}{\to} \mu(A)$, i.e., for every $U, V \in \tau$ such that $\mu(A) \subset U$ and $\mu(A) \cap V \neq \emptyset$, there exist $K \in \mathcal{K} \cap \mathcal{C}$ and $D \in \mathcal{D} \cap \mathcal{C}$, $K \subset A \subset D$ so that for every $B \in \mathcal{C}$, with $K \subset B \subset D$, we have $\mu(B) \subset U$ and $\mu(B) \cap V \neq \emptyset$).

(II) μ is said to be *regular on a subfamily* $\mathcal{S} \subseteq \mathcal{C}$ in any sense (i)–(ix) of Definition 7.4.3 if the same is every set $A \in \mathcal{S}$ with respect to μ.

In the sequel, we present some relationships among these types of regularity. Using the monotonicity of μ, one can check immediately the following:

Remark 7.4.4

(1) (i) By Remark 7.4.2, μ is always R_r^--regular. Also, if $\mathcal{C}$ is $\mathcal{B}$ or $\mathcal{B}'$, then μ is always R_l^+-regular.
(ii) Every $D \in \mathcal{D} \cap \mathcal{C}$ is R_r^+-regular and R_r^V-regular and every $K \in \mathcal{K} \cap \mathcal{C}$ is R_l^--regular and R_l^V-regular.

(2) (i) Since $\widehat{\tau}_V = \widehat{\tau}_V^+ \cup \widehat{\tau}_V^-$ and μ is always R_r^--regular, then μ is R_r^+-regular if and only if it is R_r^V-regular (if and only if for every $U \in \tau$, with $\mu(A) \subset U$, there exists $D \in \mathcal{D} \cap \mathcal{C}$ so that $A \subset D$ and $\mu(D) \subset U$).

(ii) Because μ is always R_r^--regular and $\tilde{\tau} = \tilde{\tau}_l \cap \tilde{\tau}_r$, then μ is R^--regular if and only if it is R_l^--regular.

If $\mathcal{C}$ is $\mathcal{B}$ or $\mathcal{B}'$, then, since $\tilde{\tau} = \tilde{\tau}_l \cap \tilde{\tau}_r$ and μ is always R_l^+-regular, then μ is R^+-regular if and only if it is R_r^+-regular.

(iii) If $\mathcal{C}$ is $\mathcal{B}$ or $\mathcal{B}'$, then, because $\widehat{\tau}_V = \widehat{\tau}_V^+ \cup \widehat{\tau}_V^-$ and μ is always R_l^+-regular, then μ is R_l^--regular if and only if μ is R_l^V-regular (if and only if for every $V \in \tau$ with $\mu(A) \cap V \neq \emptyset$, there exists $K \in \mathcal{K} \cap \mathcal{C}$ such that $K \subset A$ and $\mu(K) \cap V \neq \emptyset$).

(iv) μ is R^V-regular if and only if for every $U, V \in \tau$ such that $\mu(A) \subset U$ and $\mu(A) \cap V \neq \emptyset$, there exist $K \in \mathcal{K} \cap \mathcal{C}$ and $D \in \mathcal{D} \cap \mathcal{C}$, $K \subset A \subset D$ so that $\mu(D) \subset U$ and $\mu(K) \cap V \neq \emptyset$.

By the above statements, we observe that, because of the monotonicity of μ, regularity can also be interpreted as an approximation property.

(3) (i) Since $\tilde{\tau} = \tilde{\tau}_l \cap \tilde{\tau}_r$, then μ is R^V-regular if and only if it is R_r^V-regular and R_l^V-regular.

(ii) Since $\widehat{\tau}_V = \widehat{\tau}_V^+ \cup \widehat{\tau}_V^-$, then μ is R^V-regular if and only if it is R^+-regular and R^--regular.

The following result presents several relationships between regularity and increasing/decreasing convergence.

Proposition 7.4.5

(i) If $\mathcal{C}$ is $\mathcal{B}_0$ or $\mathcal{B}$ and μ is R_r^+-regular, then μ is $\widehat{\tau}_V^+$-decreasing convergent on $\mathcal{K} \cap \mathcal{C}$.

(ii) If $\mathcal{C}$ is $\mathcal{B}_0'$ or $\mathcal{B}'$ and μ is R_l^--regular, then μ is $\widehat{\tau}_V^-$-increasing convergent on $\mathcal{D} \cap \mathcal{C}$.

Proof

(i) Consider arbitrary $U \in \tau$ and $(K_n)_n, K \subset \mathcal{K} \cap \mathcal{C}$ so that $K_n \searrow K$, with $\mu(K) \subset U$.

Since μ is R_r^+-regular, by Remark 7.4.2(i), there exists $D \in \mathcal{D} \cap \mathcal{C}$, $K \subset D$ so that $\mu(D) \subset U$.

We prove there exists $n_0 \in \mathbb{N}$ so that $K \subset K_{n_0} \subset D$. For this, suppose that $K_n \cap cD \neq \emptyset$, for every $n \in \mathbb{N}$. Since $K_n \cap cD$ is compact, for every $n \in \mathbb{N}$ and $\bigcap_{i=1}^{p} K_i \cap cD = K_p \cap cD \neq \emptyset$, for every $n \in \mathbb{N}$, then $\bigcap_{i=1}^{\infty} K_i \cap cD \neq \emptyset$.

Consequently, $\emptyset \neq \bigcap_{i=1}^{\infty} K_i \cap cD = K \cap cD = \emptyset$, which is a contradiction.

Then, for every $n \geq n_0$, $\mu(K_n) \subset \mu(K_{n_0}) \subset \mu(D) \subset U$, which says that μ is decreasing convergent on $\mathcal{K} \cap \mathcal{C}$.

(ii) Consider arbitrary $U \in \tau$ and $(D_n)_n, D \subset \mathcal{D} \cap \mathcal{C}$ so that $D_n \nearrow D$, with $\mu(D) \cap U \neq \emptyset$.

Since μ is R_l^--regular, by Remark 7.4.2(iii), there exists $K \in \mathcal{K} \cap \mathcal{C}$, $K \subset D$ so that $\mu(K) \cap U \neq \emptyset$. Since $K \subset D = \bigcup_{n=1}^{\infty} D_n$, there is $n_0 \in \mathbb{N}$ so that $K \subset D_{n_0}$.

Consequently, $\mu(D_{n_0}) \cap U \neq \emptyset$, so, for every $n \geq n_0$, $\mu(D_n) \cap U \neq \emptyset$, which means that μ is increasing convergent on $\mathcal{D} \cap \mathcal{C}$.

□

Theorem 7.4.6

(i) If $\mu_1, \mu_2 : \mathcal{C} \to \mathcal{P}_0(X)$ are R_r^+-regular, then $\mu_1 = \mu_2$ on $\mathcal{D} \cap \mathcal{C}$ if and only if $\mu_1 = \mu_2$ on $\mathcal{C}$.

(ii) If μ_1, μ_2 are R_l^--regular, then $\mu_1 = \mu_2$ on $\mathcal{K} \cap \mathcal{C}$ if and only if $\mu_1 = \mu_2$ on $\mathcal{C}$.

Proof Since X is a Hausdorff, linear topological space, then $\bigcap_{V \in \mathcal{V}(0)} V = \{0\}$ and, consequently, $\bigcap_{D_0 \in \tau, 0 \in D_0} D_0 = \{0\}$. Consider arbitrary $A \in \mathcal{C}$.

(i) *The Only if part* is straightforward.

The *If part.* Let μ_1, μ_2 be R_r^+-regular, with $\mu_1 = \mu_2$ on $\mathcal{D} \cap \mathcal{C}$.

For every $D_0 \in \tau$, with $0 \in D_0$, since $\mu_1(A) \subset \mu_1(A)+D_0$, $\mu_1(A)+D_0 \in \tau$ and μ_1 is R_r^+-regular, there exists $D \in \mathcal{D} \cap \mathcal{C}$ so that $A \subset D$ and $\mu_1(D) \subset \mu_1(A) + D_0$.

Because $\mu_1(D) = \mu_2(D)$, we have $\mu_2(A) \subset \mu_2(D) \subset \mu_1(A) + D_0$. Since $D_0 \in \tau$, with $0 \in D_0$ is arbitrary, then

$$\mu_2(A) \subset \mu_1(A) + \bigcap_{D_0 \in \tau, 0 \in D_0} D_0 = \mu_1(A) + \{0\} = \mu_1(A).$$

Analogously, $\mu_1(A) \subset \mu_2(A)$, so, finally, $\mu_1(A) = \mu_2(A)$.

(ii) *The Only if part* is straightforward.

The *If part.* Let μ_1, μ_2 be R_l^--regular, with $\mu_1 = \mu_2$ on $\mathcal{K} \cap \mathcal{C}$.

For every $x \in \mu_1(A)$ and arbitrary, fixed $D_0 \in \tau$, with $0 \in D_0$, since $\mu_1(A) \cap [\{x\} + D_0] \neq \emptyset$, $\{x\} + D_0 \in \tau$ and μ_1 is R_l^--regular, there exists $K \in \mathcal{K} \cap \mathcal{C}$ so that $K \subset A$ and $\mu_1(K) \cap [\{x\} + D_0] \neq \emptyset$.

Because $\mu_1(K) = \mu_2(K)$, we get $\mu_2(A) \cap [\{x\} + D_0] \neq \emptyset$, so $x \in \mu_2(A) - D_0$. Consequently, $\mu_1(A) \subset \mu_2(A) - D_0$. Since $D_0 \in \tau$, with $0 \in D_0$ is arbitrary, then

$$\mu_1(A) \subset \mu_2(A) + \left[\bigcap_{D_0 \in \tau, 0 \in D_0} - D_0 \right] = \mu_2(A),$$

whence $x \in \mu_2(A)$, which implies that $\mu_1(A) \subset \mu_2(A)$.

Analogously, $\mu_2(A) \subset \mu_1(A)$, so $\mu_1(A) = \mu_2(A)$.

□

Using the monotonicity of μ, we also give the following notions:

Definition 7.4.7

(I) $A \in \mathcal{C}$ is said to be (with respect to μ):

(i) $R_l^{\prime +}$-*regular* if for every $V_0 \in \mathcal{V}(0)$, there is $K \in \mathcal{K} \cap \mathcal{C}$ so that $K \subset A$ and $\mu(A \backslash K) \subset V_0$.

(ii) $R_r'^+$*-regular* if for every $V_0 \in \mathcal{V}(0)$, there is $D \in \mathcal{D} \cap \mathcal{C}$ so that $A \subset D$ and $\mu(D \backslash A) \subset V_0$.
(iii) R'^+*-regular* if for every $V_0 \in \mathcal{V}(0)$, there are $K \in \mathcal{K} \cap \mathcal{C}$ and $D \in \mathcal{D} \cap \mathcal{C}$ such that $K \subset A \subset D$ and $\mu(D \backslash K) \subset V_0$.

(II) μ is said to be *regular* in any sense (i)–(iii) from above if the same is every set $A \in \mathcal{C}$ with respect to μ.

Obviously, every $K \in \mathcal{K} \cap \mathcal{C}$ is $R_l'^+$-regular and every $D \in \mathcal{D} \cap \mathcal{C}$ is $R_r'^+$-regular. In what follows, we establish relationships among $R_l'^+$-regularity, $R_r'^+$-regularity and R'^+-regularity.

Proposition 7.4.8

(i) If μ is R'^+-regular, then μ is $R_l'^+$-regular and $R_r'^+$-regular.
(ii) If μ is a multisubmeasure, then μ is R'^+-regular if and only if it is $R_l'^+$-regular and $R_r'^+$-regular.
(iii) Suppose T is compact and $\mu : \mathcal{B} \to \mathcal{P}_0(X)$ is a $R_l'^+$-regular (or, respectively, $R_r'^+$-regular) multisubmeasure. Then μ is R'^+-regular.
Consequently, if T is compact and $\mu : \mathcal{B} \to \mathcal{P}_0(X)$ is a multisubmeasure, then $R_l'^+$-regularity, $R_r'^+$-regularity and R'^+-regularity are equivalent.

Proof

(i) The statement is straightforward due to the monotonicity of μ.
(ii) Consider arbitrary $A \in \mathcal{C}$. Since X is a Hausdorff, linear topological space, then for every $V_0 \in \mathcal{V}(0)$, there exists $W_0 \in \mathcal{V}(0)$ so that $W_0 + W_0 \subset V_0$. If μ is $R_l'^+$-regular and $R_r'^+$-regular, there are $K \in \mathcal{K} \cap \mathcal{C}$ and $D \in \mathcal{D} \cap \mathcal{C}$ such that $K \subset A \subset D$, $\mu(A \backslash K) \subset W_0$ and $\mu(D \backslash A) \subset W_0$. Consequently, since μ is a multisubmeasure, we have

$$\mu(D \backslash K) \subset \mu(D \backslash A) + \mu(A \backslash K) \subset W_0 + W_0 \subset V_0,$$

so μ is R'^+-regular.
Now, the conclusion follows by (i).
(iii) Evidently, since T is compact, then $\mathcal{C} = \mathcal{B}$ is an algebra. Consider arbitrary $A \in \mathcal{C}$. For every $V_0 \in \mathcal{V}(0)$, there exists $W_0 \in \mathcal{V}(0)$ so that $W_0 + W_0 \subset V_0$. If μ is $R_l'^+$-regular, there exists $K \in \mathcal{K} \cap \mathcal{C}$ so that $K \subset A$ and $\mu(A \backslash K) \subset W_0$.

Analogously, for $cA \in \mathcal{C}$, there exists $\widetilde{K} \in \mathcal{K} \cap \mathcal{C}$ so that $\widetilde{K} \subset cA$ and $\mu(cA \backslash \widetilde{K}) \subset W_0$.
If we denote $D = c\widetilde{K}$, then $D \in \mathcal{D} \cap \mathcal{C}$, $A \subset D$ and $\mu(D \backslash A) \subset W_0$.
Since μ is a multisubmeasure, we have

$$\mu(D \backslash K) \subset \mu(D \backslash A) + \mu(A \backslash K) \subset W_0 + W_0 \subset V_0,$$

that is, μ is R'^+-regular.

Taking into account that T is a compact space, one can use similar argues in case when μ is $R_r'^{+}$-regular. □

Definition 7.4.9

(i) A set $A \in \mathcal{C}$ is said to be an *atom* of μ if $\mu(A) \supsetneq \{0\}$ and for every $B \in \mathcal{C}$, with $B \subset A$, we have $\mu(B) = \{0\}$ or $\mu(A \backslash B) = \{0\}$.

(ii) μ is said to be *non-atomic* if it has no atoms (i.e., for every $A \in \mathcal{C}$, with $\mu(A) \supsetneq \{0\}$, there exists $B \in \mathcal{C}$, with $B \subset A$, $\mu(B) \supsetneq \{0\}$ and $\mu(A \backslash B) \supsetneq \{0\}$).

In a similar manner as in [7], we can prove the following result, which enables us to characterize atoms and non-atomicity of a $R_l'^{+}$-regular multisubmeasure.

Lemma 7.4.10 *If $\mu : \mathcal{B} \to \mathcal{P}_0(X)$ is a $R_l'^{+}$-regular multisubmeasure, then:*

(i) a set $A \in \mathcal{B}$, with $\mu(A) \supsetneq \{0\}$ is an atom of μ if and only if $\exists! a \in A$ so that $\mu(A \backslash \{a\}) = \{0\}$.

(ii) μ is non-atomic if and only if for every $t \in T$, $\mu(\{t\}) = \{0\}$.

Proof Let $A \in \mathcal{B}$ be an atom of μ and $\mathcal{K}_A = \{K \subseteq A;\ K \in \mathcal{K} \text{ and } \mu(A \backslash K) = \{0\}\}$.

We shall only prove that $\mathcal{K}_A$ is nonvoid (the rest of the proof is similar to that from [7]).

On the contrary, we suppose that $\mathcal{K}_A = \emptyset$. So, for every $K \subseteq A$, with $K \in \mathcal{K}$, we have $\mu(A \backslash K) \supsetneq \{0\}$. Since A is an atom, then $\mu(K) = \{0\}$.

On the other hand, because μ is $R_l'^{+}$-regular, then for every $V_0 \in \mathcal{V}(0)$, there is $K_0 \in \mathcal{K}$ so that $K_0 \subset A$ and $\mu(A \backslash K_0) \subset V_0$.

Consequently, $\mu(A) \subset \mu(A \backslash K_0) + \mu(K) \subset V_0$, for every $V_0 \in \mathcal{V}(0)$, whence $\mu(A) \subset \bigcap_{V_0 \in \mathcal{V}(0)} V_0 = \{0\}$, so $\mu(A) = \{0\}$, which is a contradiction. □

Now, we establish a set-valued Alexandroff type theorem in Vietoris topology:

Theorem 7.4.11 (Alexandroff theorem) *Suppose X is a metrizable, linear topological space. If $\mu : \mathcal{C} \to \mathcal{P}_0(X)$ is a $R_l'^{+}$-regular multisubmeasure, then μ is $\widehat{\tau}_V^{+}$-order continuous.*

Proof Since X is a metrizable, linear topological space, by Precupanu [17] there exists a countable fundamental system of neighbourhoods of the origin, $\mathcal{U}(0) = \{V_n\}_{n \in \mathbb{N}}$ so that, for every arbitrary fixed $n \in \mathbb{N}$,

$$(*)\ V_{n+1} + V_{n+2} + \cdots + V_{n+p} \subset V_n, \text{ for every } p \in \mathbb{N}^*.$$

Let be arbitrary $V \in \mathcal{V}(0)$. There exists $\widetilde{V} \in \mathcal{U}(0)$ so that $\widetilde{V} \subset V$. Consider arbitrary decreasing sequence $(A_n)_{n \in \mathbb{N}} \subset \mathcal{C}$, with $A_n \searrow \emptyset$. Evidently, there exists $n_0 \in \mathbb{N}$ so that $\widetilde{V} = V_{n_0}$ and by $(*)$, for every $p \in \mathbb{N}^*$,

$$V_{n_0+1} + V_{n_0+2} + \cdots + V_{n_0+p} \subset \widetilde{V}.$$

Since μ is $R_l'^+$-regular, for every $i \in \mathbb{N}^*$, there is $K_i \in \mathcal{K} \cap \mathcal{C}$ so that $K_i \subset A_i$ and $\mu(A_i \backslash K_i) \subset V_{n_0+i}$.

Denote $K_1' = K_1$. Then $K_1' \in \mathcal{K} \cap \mathcal{C}$, $K_1' \subset A_1$ and

$$\mu(A_1 \backslash K_1') = \mu(A_1 \backslash K_1) \subset V_{n_0+1} \subset \widetilde{V}.$$

Denote $K_2' = K_1 \cap K_2$. Then $K_2' \in \mathcal{K} \cap \mathcal{C}$, $K_2' \subset A_2$ and

$$\mu(A_2 \backslash K_2') \subset \mu(A_1 \backslash K_1) + \mu(A_2 \backslash K_2) \subset V_{n_0+1} + V_{n_0+2} \subset \widetilde{V}.$$

Suppose we have recurrently constructed $K_1', K_2', \dots, K_{p-1}'$ and let $K_p' = K_1 \cap K_2 \cap \dots \cap K_p$. Then $K_p' \in \mathcal{K} \cap \mathcal{C}$, $K_p' \subset A_p$, $K_{p-1}' \supset K_p'$ and

$$\mu(A_p \backslash K_p') \subset \mu(A_1 \backslash K_1) + \dots + \mu(A_p \backslash K_p) \subset V_{n_0+1} + V_{n_0+2} + \dots + V_{n_0+p} \subset \widetilde{V}.$$

So, we have recurrently constructed a decreasing sequence $(K_p')_p \subset \mathcal{K} \cap \mathcal{C}$ so that for every $p \in \mathbb{N}^*$, $K_p' \subset A_p$, $K_p' \supset K_{p+1}'$ and $\mu(A_p \backslash K_p') \subset \widetilde{V}$.

Since $K_n' \subset A_n$, for every $n \in \mathbb{N}$ and $\bigcap\limits_{n=1}^{\infty} A_n = \emptyset$, then $\bigcap\limits_{n=1}^{\infty} K_n' = \emptyset$. Because $K_n' \subset K_1'$, for every $n \in \mathbb{N}^*$ and K_1' is compact, there exists $n_0'(\varepsilon) \in \mathbb{N}^*$ so that $\bigcap\limits_{i=1}^{n_0'} K_i' = K_{n_0'}' = \emptyset$. Consequently, for every $n \geq n_0'$, $\mu(A_n) \subset \mu(A_{n_0'} \backslash K_{n_0'}') \subset \widetilde{V} \subset V$, whence μ is $\widehat{\tau}_V^+$-order continuous. □

In what follows, we establish a converse of Theorem 7.4.11 for the special case when $\mathcal{C} = \mathcal{B}_0$. First, we need the following lemma:

Lemma 7.4.12 (Operations with R'^+-regular sets) *Suppose $\mu : \mathcal{C} \to \mathcal{P}_0(X)$ is a multisubmeasure.*

(i) If A_1, A_2 are R'^+-regular, then $A_1 \backslash A_2$ is R'^+-regular;

(ii) If $p \in \mathbb{N}^*$ is arbitrary, fixed and if $(A_i)_{i=\overline{1,p}}$ are R'^+-regular, then $A = \bigcup\limits_{i=1}^{p} A_i$ is R'^+-regular;

(iii) If X is a metrizable, linear topological space, $\mathcal{C}$ is a δ-ring, A_r is R'^+-regular, for every $r \in \mathbb{N}$ and $A_r \searrow A$, then A is R'^+-regular.

Proof

(i) Consider arbitrary R'^+-regular sets A_1 and A_2 and let $U_0 \in \mathcal{V}(0)$.

Since X is a Hausdorff, linear topological space, there exists $W_0 \in \mathcal{V}(0)$ so that $W_0 + W_0 \subset U_0$.

Because A_1 and A_2 are R'^+-regular, there exist $K_1, K_2 \in \mathcal{K} \cap \mathcal{C}$ and $D_1, D_2 \in \mathcal{D} \cap \mathcal{C}$ so that $K_1 \subset A_1 \subset D_1$, $K_2 \subset A_2 \subset D_2$, $\mu(D_1 \backslash K_1) \subset W_0$ and $\mu(D_2 \backslash K_2) \subset W_0$.

We observe that $D_1\backslash K_2 \in \mathcal{D}\cap\mathcal{C}$, $K_1\backslash D_2 \in \mathcal{K}\cap\mathcal{C}$ and $K_1\backslash D_2 \subset A_1\backslash A_2 \subset D_1\backslash K_2$. Since

$$\begin{aligned}\mu((D_1\backslash K_2)\backslash(K_1\backslash D_2)) &\subset \mu((D_1\backslash K_1)\cup(D_2\backslash K_2))\\ &\subset \mu(D_1\backslash K_1)+\mu(D_2\backslash K_2)\subset\\ &\subset W_0+W_0\subset U_0,\end{aligned}$$

then $A_1\backslash A_2$ is R'^{+}-regular.

(ii) Suppose $p \in \mathbb{N}^*$ is arbitrary, fixed and consider arbitrary R'^{+}-regular sets $(A_i)_{i=\overline{1,p}}$. Let $U_0 \in \mathcal{V}(0)$.

Since X is a Hausdorff linear topological space, there exists $V_p \in \mathcal{V}(0)$ so that $\underbrace{V_p + V_p + \cdots + V_p}_{p} \subset U_0$.

Because for every $i = \overline{1, p}$, A_i is R'^{+}-regular, there exists $K_i \in \mathcal{K}\cap\mathcal{C}$ and $D_i \in \mathcal{D}\cap\mathcal{C}$ so that $K_i \subset A_i \subset D_i$ and $\mu(D_i\backslash K_i) \subset V_p$.

We observe that $\bigcup\limits_{i=1}^{p} D_i \in \mathcal{D}\cap\mathcal{C}$, $\bigcup\limits_{i=1}^{p} K_i \in \mathcal{K}\cap\mathcal{C}$ and $\bigcup\limits_{i=1}^{p} K_i \subset \bigcup\limits_{i=1}^{p} A_i \subset \bigcup\limits_{i=1}^{p} D_i$. Since

$$\begin{aligned}\mu\left(\left(\bigcup_{i=1}^{p} D_i\right)\backslash\left(\bigcup_{i=1}^{p} K_i\right)\right) &\subset \mu\left(\bigcup_{i=1}^{p}(D_i\backslash K_i)\right)\subset\\ &\subset \mu(D_1\backslash K_1)+\mu(D_2\backslash K_2)+\cdots+\mu(D_p\backslash K_p)\subset\\ &\subset V_p+V_p+\cdots+V_p\subset U_0,\end{aligned}$$

we get that $\bigcup\limits_{i=1}^{p} A_i$ is R'^{+}-regular.

(iii) Since X is a metrizable, linear topological space, there exists a fundamental system of neighbourhoods of the origin, $\mathcal{U}(0) = \{V_n\}_{n\in\mathbb{N}}$ so that, for every arbitrary fixed $n \in \mathbb{N}$, $(*)$ holds.

Consider arbitrary $V \in \mathcal{V}(0)$. There exists $\widetilde{V} \in \mathcal{U}(0)$ so that $\widetilde{V} \subset V$. Obviously, there is $i_0 \in \mathbb{N}$ so that $\widetilde{V} = V_{i_0}$ and for every $p \in \mathbb{N}^*$,

$$V_{i_0+1} + V_{i_0+2} + \cdots + V_{i_0+p} \subset \widetilde{V}.$$

Since for every $r \in \mathbb{N}$, A_r is R'^{+}-regular, there exist $K_r \in \mathcal{K}\cap\mathcal{C}$ and $D_r \in \mathcal{D}\cap\mathcal{C}$ so that $K_r \subset A_r \subset D_r$ and $\mu(D_r\backslash K_r) \subset V_{i_0+r+1}$.

Denote $C_r = \bigcap\limits_{i=1}^{r} K_i$, $E_r = \bigcap\limits_{i=1}^{r} D_i$ and $C = \bigcap\limits_{r=1}^{\infty} C_r$. Evidently, for every $r \in \mathbb{N}$, $C_r, C \in \mathcal{K}\cap\mathcal{C}$ and $E_r \in \mathcal{D}\cap\mathcal{C}$.

Since $C_r \searrow C$ and μ is $\widehat{\tau}_V^{+}$-order continuous, there exists $r_0 \in \mathbb{N}^*$ so that $\mu(C_{r_0}\backslash C) \subset V_{i_0+1}$.

We observe that $C \subset C_{r_0} \subset A_{r_0} \subset E_{r_0}, C \subset A \subset A_{r_0} \subset E_{r_0}$ and $E_{r_0}\backslash C_{r_0} = (\bigcap_{i=1}^{r_0} D_i)\backslash(\bigcap_{i=1}^{r_0} K_i) \subset \bigcup_{i=1}^{r_0}(D_i\backslash K_i)$, so

$$\mu(E_{r_0}\backslash C_{r_0}) \subset \mu(D_1\backslash K_1) + \cdots + \mu(D_{r_0}\backslash K_{r_0}) \subset V_{i_0+2} + \cdots + V_{i_0+r_0+1}.$$

Then

$$\mu(E_{r_0}\backslash C) = \mu((E_{r_0}\backslash C_{r_0}) \cup (C_{r_0}\backslash C)) \subset$$
$$\subset V_{i_0+1} + V_{i_0+2} + \cdots + V_{i_0+r_0+1} \subset \widetilde{V} \subset V,$$

so, A is R'^{+}-regular.

□

Theorem 7.4.13 *If X is a metrizable, linear topological space and if $\mu : \mathcal{B}_0 \to \mathcal{P}_0(X)$ is a $\widehat{\tau}_V^+$-order continuous multisubmeasure, then μ is R'^{+}-regular.*

Proof Denote $\mathcal{A} = \{A \in \mathcal{B}_0 / A$ is R'^{+}-regular with respect to $\mu\}$.

We shall prove that $\mathcal{A} = \mathcal{B}_0$. First, we observe that $\mathcal{A}$ is nonvoid. Indeed, by Dinculeanu [4, Ch. III, p. 190], for every G_δ, compact set $K \subset T$, there exists a decreasing sequence of sets $(D_n)_n \subset \mathcal{D} \cap \mathcal{B}_0$ so that $D_n \searrow K$. Since μ is $\widehat{\tau}_V^+$-order continuous, for every $U_0 \in \mathcal{V}(0)$, there exists $n_0 \in \mathbb{N}$ so that $\mu(D_{n_0}\backslash K) \subset U_0$, that is, K is R'^{+}-regular. Consequently, $\mathcal{A}$ contains the family of all G_δ, compact sets of T, so $\mathcal{A}$ is nonvoid.

By Lemma 7.4.10, $\mathcal{A}$ is a δ-ring which contains all G_δ, compact sets of T. Since $\mathcal{B}_0$ is the δ-ring generated by G_δ, compact sets of T, then $\mathcal{A} = \mathcal{B}_0$, so μ is R'^{+}-regular. □

By Proposition 7.4.8(i), Theorems 7.4.11 and 7.4.13, we immediately have:

Corollary 7.4.14 *If X is a metrizable, linear topological space, then a multisubmeasure $\mu : \mathcal{B}_0 \to \mathcal{P}_0(X)$ is $\widehat{\tau}_V^+$-order continuous if and only if μ is $R_l'^{+}$-regular.*

In what follows, we compare the types of regularity in Vietoris topology $\widehat{\tau}_V$ for $\mathcal{P}_0(X)$-valued monotone set multifunctions (where X is a linear topological Hausdorff space), with those we studied in [6–9] in Hausdorff topology $\widehat{\tau}_H$ for $\mathcal{P}_c(X)$-valued monotone set multifunctions (where X was a real normed space). Refer to [6–9] for notions, notations and results concerning regularity in Hausdorff topology.

As we shall see in the following, although for $\mathcal{P}_k(X)$-valued monotone set multifunctions (X being a real normed space), since $\hat{\tau}_H^- = \hat{\tau}_V^-$, $\hat{\tau}_H^+ = \hat{\tau}_V^+$ and $\hat{\tau}_H = \hat{\tau}_V$, we have the equivalence between regularity properties in Hausdorff topology, respectively, in Vietoris topology, we shall demonstrate that several similar corresponding results hold even if μ is only $\mathcal{P}_c(X)$-valued.

Suppose X **is a real normed space** and $\mu : \mathcal{C} \to \mathcal{P}_c(X)$ is **a monotone set multifunction**.

Definition 7.4.15

(I) An arbitrary set $A \in \mathcal{C}$ is said to be:

(i) R^H*-regular* if $\mu : (\mathcal{C}, \tilde{\tau}) \to (\mathcal{P}_c(X), \widehat{\tau}_H)$ is continuous at A (i.e., $B \xrightarrow{\tilde{\tau}} A \Rightarrow \mu(B) \xrightarrow{\widehat{\tau}_H} \mu(A)$, i.e., for every $\varepsilon > 0$, there are $K \in \mathcal{K} \cap \mathcal{C}$, $K \subset A$ and $D \in \mathcal{D} \cap \mathcal{C}$, $D \supset A$ so that $h(\mu(A), \mu(B)) < \varepsilon$, for every $B \in \mathcal{C}$, with $K \subset B \subset D$).

(ii) R_l^H*-regular* if $\mu : (\mathcal{C}, \tilde{\tau}_l) \to (\mathcal{P}_c(X), \widehat{\tau}_H)$ is continuous at A (i.e., $B \xrightarrow{\tilde{\tau}_l} A \Rightarrow \mu(B) \xrightarrow{\widehat{\tau}_H} \mu(A)$, i.e., for every $\varepsilon > 0$, there exists $K \in \mathcal{K} \cap \mathcal{C}$, $K \subset A$ so that $h(\mu(A), \mu(B)) < \varepsilon$, for every $B \in \mathcal{C}$, with $K \subset B \subset A$).

(iii) R_r^H*-regular* if $\mu : (\mathcal{C}, \tilde{\tau}_r) \to (\mathcal{P}_c(X), \widehat{\tau}_H)$ is continuous at A (i.e., $B \xrightarrow{\tilde{\tau}_r} A \Rightarrow \mu(B) \xrightarrow{\widehat{\tau}_H} \mu(A)$, i.e., for every $\varepsilon > 0$, there exists $D \in \mathcal{D} \cap \mathcal{C}$, $D \supset A$ such that $h(\mu(A), \mu(B)) < \varepsilon$, for every $B \in \mathcal{C}$, with $A \subset B \subset D$).

(II) μ is said to be:

(i) R^H*-regular* (respectively, R_l^H*-regular*, R_r^H*-regular*) if every set $A \in \mathcal{C}$ is R^H-regular (respectively, R_l^H-regular, R_r^H-regular).

Remark 7.4.16

(1) Due to the monotonicity of μ, we can easily observe that:

(i) μ is R^H-regular if and only if for every $\varepsilon > 0$, there are $K \in \mathcal{K} \cap \mathcal{C}$, $K \subset A$ and $D \in \mathcal{D} \cap \mathcal{C}$, $D \supset A$ so that $e(\mu(D), \mu(K)) < \varepsilon$;

(ii) μ is R_l^H-regular if and only if for every $\varepsilon > 0$, there exists $K \in \mathcal{K} \cap \mathcal{C}$, $K \subset A$ so that $e(\mu(A), \mu(K)) < \varepsilon$;

(iii) μ is R_r^H-regular if and only if for every $\varepsilon > 0$, there exists $D \in \mathcal{D} \cap \mathcal{C}$, $D \supset A$ such that $e(\mu(D), \mu(A)) < \varepsilon$,

which are exactly the types of regularity (as an approximation property) we introduced and studied in [6–9] in Hausdorff topology.

(2) In [6, 9], we have proved that μ is R^H-regular if and only if it is R_l^H-regular and R_r^H-regular. In fact, we observe that the statement is straightforward, since $\tilde{\tau} = \tilde{\tau}_l \cap \tilde{\tau}_r$, interpreting regularity as a continuity property and using (1).

(3) (i) μ is $R_l^{\prime +}$-regular if and only if it is $R_l^{\prime}$-regular (in the sense of [6–9]).
(ii) μ is $R_r^{\prime +}$-regular if and only if it is $R_r^{\prime}$-regular (in the sense of [6–9]).
(iii) μ is $R^{\prime +}$-regular if and only if it is $R^{\prime}$-regular (in the sense of [6–9]).

Theorem 7.4.17

(i) If μ is R_r^+-regular, then μ is R_r^H-regular.
(ii) If μ is R_l^H-regular, then μ is R_l^--regular.
(iii) If $\mu : \mathcal{C} \to \mathcal{P}_k(X)$, then:

(a) μ is R_r^+-regular if and only if μ is R_r^H-regular.
(b) μ is R_l^--regular if and only if μ is R_l^H-regular.

Proof

(i) Consider arbitrary $A \in \mathcal{C}$ and $\varepsilon > 0$. Since $\mu(A) \subset \bigcup_{x\in\mu(A)} S(x,\varepsilon) = U, U \in \tau$ and μ is R_r^+-regular, there exists $D \in \mathcal{D} \cap \mathcal{C}$ so that $A \subset D$ and $\mu(D) \subset \bigcup_{x\in\mu(A)} S\big(x, \frac{\varepsilon}{2}\big)$, so, for every $y \in \mu(D)$, there exists $x_y \in \mu(A)$ so that $d(y, x_y) < \frac{\varepsilon}{2}$, whence $d(y, \mu(A)) < \frac{\varepsilon}{2}$. Consequently, $e(\mu(D), \mu(A)) < \varepsilon$, that is, according to Remark 7.4.16-(1)(iii), μ is R_r^H-regular.

(ii) Consider arbitrary $A \in \mathcal{C}$ and $V \in \tau$, with $\mu(A) \cap V \neq \emptyset$. There is $x_0 \in \mu(A)$ and $\varepsilon_0 > 0$ such that $S(x_0, \varepsilon_0) \subset V$. Since μ is R_l^H-regular, by Remark 7.4.16-(1)(ii), for ε_0 there is $K \in \mathcal{K} \cap \mathcal{C}$, so that $K \subset A$ and $e(\mu(A), \mu(K)) < \varepsilon_0$.

Consequently, $d(x_0, \mu(K)) < \varepsilon_0$, so there exists $y_0 \in \mu(K)$ such that $d(x_0, y_0) < \varepsilon_0$. Therefore, $\mu(K) \cap S(x_0, \varepsilon_0) \neq \emptyset$, which implies $\mu(K) \cap V \neq \emptyset$, so, μ is R_l^--regular.

(iii) The statements follow because μ is monotone and $\widehat{\tau}_V^+ = \widehat{\tau}_H^+$ and $\widehat{\tau}_V^- = \widehat{\tau}_H^-$ on $\mathcal{P}_k(X)$.

□

7.5 Conclusions

In this chapter, continuity properties, such as regularity, increasing/decreasing convergence, order continuity and exhaustivity, are studied in Vietoris topology for $\mathcal{P}_0(X)$-valued monotone set multifunctions, X being a Hausdorff, linear topological space. An Alexandroff type theorem in Vietoris topology and its converse are also established.

Then, for the particular case when X is a real normed space, these continuity properties are compared to corresponding properties in Hausdorff topology for $\mathcal{P}_c(X)$-valued monotone set multifunctions.

Thus, the present chapter introduces generalizations of important continuity properties coming from fuzzy (i.e., monotone) Measure Theory. Precisely, these generalizations are firstly made for $\mathcal{P}_c(X)$-valued monotone set multifunctions (X being a real normed space), and secondly, for $\mathcal{P}_0(X)$-valued monotone set multifunctions (X being, more generally, a Hausdorff linear topological space).

Based on these results, in Chap. 8 we shall continue our investigations, obtaining Egoroff and Lusin type theorems in Vietoris topology.

References

1. Apreutesei, G.: Families of subsets and the coincidence of hypertopologies. An. Şt. Univ. Iaşi XLIX, 1–18 (2003)
2. Beer, G.: Topologies on Closed and Closed Convex Sets. Kluwer, Dordrecht (1993)
3. Billara, V.: Topologies for 2^X. Macmillan, New York (1963)

4. Dinculeanu, N.: Measure Theory and Real Functions (in Romanian). Ed. Did. şi Ped., Bucureşti (1964)
5. Gavriluţ, A.: Properties of regularity for multisubmeasures with respect to the Vietoris topology. An. Şt. Univ. Iaşi, L, s. I a f. **2**, 373–392 (2004)
6. Gavriluţ, A.: Properties of Regularity for Set Multifunctions (in Romanian). Venus Publishing House, Iaşi (2006)
7. Gavriluţ, A.: Non-atomicity and the Darboux property for fuzzy and non-fuzzy Borel/Baire multivalued set functions. Fuzzy Sets Syst. **160**, 1308–1317 (2009). Erratum in Fuzzy Sets and Systems **161**, 2612–2613 (2010)
8. Gavriluţ, A.: Regularity and autocontinuity of set multifunctions. Fuzzy Sets Syst. **161**, 681–693 (2010)
9. Gavriluţ, A.: Abstract regular null-null-additive set multifunctions in Hausdorff topology, An. Şt. Univ. Iaşi **59**(1). https://doi.org/10.2478/v10157-012-0029-4
10. Guo, C., Zhang, D.: On the set-valued fuzzy measures. Inform. Sci. **160**, 13–25 (2004)
11. Hu, S., Papageorgiou, N.S.: Handbook of Multivalued Analysis, vol. I. Kluwer, Dordrecht (1997)
12. Jiang, Q., Suzuki, H.: Lebesque and Saks decompositions of σ-finite fuzzy measures. Fuzzy Sets Syst. **75**, 373–385 (1995)
13. Jiang, Q., Suzuki, H., Wang, Z., Klir, G.: Exhaustivity and absolute continuity of fuzzy measures. Fuzzy Sets Syst. **96**, 231–238 (1998)
14. Klein, E., Thompson, A.: Theory of Correspondences. Wiley, New York (1984)
15. Michael, E.: Topologies on spaces of subsets. Trans. AMS **71**, 152–182 (1951)
16. Pap, E.: Null-additive Set Functions. Kluwer, Dordrecht (1995)
17. Precupanu, T.: Linear Topological Spaces and Elements of Convex Analysis (in Romanian). Ed. Acad. Romania (1992)
18. Precupanu, A., Gavriluţ, A.: A set-valued Egoroff type theorem. Fuzzy Sets Syst. **175**, 87–95 (2011)
19. Precupanu, A., Gavriluţ, A.: Set-valued Lusin type theorem for null-null-additive set multifunctions. Fuzzy Sets Syst. **204**, 106–116 (2012)
20. Precupanu, A., Croitoru, A., Godet-Thobie, Ch.: Set-valued Integrals (in Romanian). Iaşi (2011)
21. Precupanu, A., Precupanu, T., Turinici, M., Apreutesei Dumitriu, N., Stamate, C., Satco, B.R., Văideanu, C., Apreutesei, G., Rusu, D., Gavriluţ, A.C., Apetrii, M.: Modern Directions in Multivalued Analysis and Optimization Theory (in Romanian). Venus Publishing House, Iaşi (2006)
22. Sugeno, M.: Theory of fuzzy integrals and its applications, Ph.D. Thesis, Tokyo Institute of Technology (1974)
23. Zhang, D., Guo, C.: Generalized fuzzy integrals of set-valued functions. Fuzzy Sets Syst. **76**, 365–373 (1995)
24. Zhang, D., Wang, Z.: On set-valued fuzzy integrals. Fuzzy Sets Syst. **56**, 237–241 (1993)

Chapter 8
Approximation theorems for fuzzy set multifunctions in Vietoris topology: Physical implications of regularity

8.1 Introduction

In this chapter, we continue our study began in Chap. 7, concerning continuity properties (especially, regularity, also viewed as an approximation property) for $\mathcal{P}_0(X)$-valued set multifunctions, X being a linear, topological space. The purpose is to obtain here Egoroff and Lusin type theorems for set multifunctions in Vietoris hypertopology. Some mathematical applications are established and several physical implications of the mathematical model of regularity are presented, and this fact allows a classification of the physical models.

Non-additive measures can be used for modelling problems in non-deterministic environment. In the last years, this area has been widely developed and a wide variety of topics have been investigated. Also, non-additive integrals have very interesting properties from a mathematical point of view, which have been studied and applied to various fields: information sciences, decision-making problems, monotone expectation, aggregation approach, etc.

The idea of modeling the behaviour of phenomena at multiple scales has become a useful tool in pure and applied mathematics. Fractal-based techniques lie at the heart of this area, since fractals are multiscale objects, which often describe such phenomena better than traditional mathematical models. In Kunze et al. [21] and Wicks [45], certain hyperspace theories concerning the Hausdorff metric and the Vietoris topology, as a foundation for self-similarity and fractality are developed. In fact, from a mathematical point of view, fractality has important commonalities, both with various hypertopologies (Hausdorff, Vietoris topologies [2, 3, 6, 21, 44]) and with the regularity of Borel measures.

For many years, topological methods were used in many fields to study the chaotic nature in dynamical systems (Sharma and Nagar [42], Wang et al. [44], Goméz-Rueda et al. [15], Li [23], Liu et al. [26], Ma et al. [29], Fu and Xing [10]), which seem to be collective phenomenon emerging out of many segregated components. Most of these systems are collective (set-valued) dynamics of many

A. Gavriluţ et al., *Atomicity through Fractal Measure Theory*,
https://doi.org/10.1007/978-3-030-29593-6_8

units of individual systems. It therefore arises the need of a topological treatment of such collective dynamics. Recent studies of dynamical systems, in engineering and physical sciences, have revealed that the underlying dynamics is set-valued (collective), and not of a normal, individual kind, as it was usually studied before. We mention interesting approaches of topology in psychology, in the studies [22] of Lewin et al. and [8] of Brown. We also note different considerations concerning generalized fractals in hyperspaces endowed with the Hausdorff, or, more generally, with the Vietoris hypertopology (Andres and Fiser [2], Andres and Rypka [3], Banakh and Novosad [6], Kunze et al. [21]).

Because of its numerous applications in problems of optimization, convex analysis, economy, the study of hypertopologies (for instance, Hausdorff, Vietoris, Wijsman, Fell, Attouch-Wets) has become of a great interest. Important results concerning the Vietoris topology can be found in Beer [7], Apreutesei [4], Hu and Papageorgiou [18], etc. Interesting results involving, for instance, the Hausdorff distance were obtained by Lorenzo and Maio [27] in melodic similarity, Lu et al. [28]—an approach to word image matching, etc.

Due to their various applications in economics, games theory, artificial intelligence and many other important fields, many considerations from non-additive measures theory (especially, continuity properties) were extended to the set-valued case (Guo and Zhang [16] in Kuratowski topology, Gavriluţ [11–14] and Precupanu and Gavriluţ [38, 39] in Hausdorff topology).

Regularity property is an important continuity property in different topologies, but, at the same time, it can be interpreted as an approximation property. Precisely, in this way, we can approximate no matter how well different "unknown" sets by other sets which we have more information. Usually, from a mathematical perspective, this approximation is done from the left by open, or more restrictive, by compact sets and/or from the right by open sets. In this chapter, we shall point out important mathematical and physical applications of regularity.

The classical Lusin's theorem concerning the existence of continuous restrictions of measurable functions is very important and useful for discussing different kinds of approximation of measurable functions defined on special topological spaces and for numerous applications in the study of convergence of sequences of Sugeno and Choquet integrable functions (Li and Yasuda [24]). In the last years, Lusin's result was generalized for non-additive set functions by many authors: Kawabe [20], Li and Yasuda [24], Song and Li [43], Jiang and Suzuki [19], Pap [35], etc. Their proofs are based on the one hand, on different versions of Egoroff's theorem and, on the other hand, on various types of regularity of the non-additive measures. An interesting application of the Lusin theorem is due to Li et al. [25], in the study of the approximation properties of neural networks, as the learning ability of a neural network is closely related to its approximating capabilities. Recently, Lusin's theorem was considered in the set-valued case by Averna [5] for measurable multifunctions.

In [13] we have introduced and studied different notions of continuity, including regularity, in the Vietoris hypertopology, for $\mathcal{P}_0(X)$-valued set multifunctions, X being a Hausdorff, linear topological space.

The important theoretical and practical problem of obtaining Lusin type theorems in the set-valued case in different hypertopologies has been very recently investigated. Precisely, in [11] we have proved a Lusin type theorem in the Hausdorff hypertopology for regular uniformly autocontinuous monotone set multifunctions and in Precupanu and Gavriluţ [39] we generalized these results for regular null-additive monotone set multifunctions in the same Hausdorff hypertopology.

In this chapter, based on the results established in the previous chapter and also in [13], we continue our investigations concerning regularity in the set-valued setting of Vietoris hypertopology. Namely, we are now able to establish set-valued Egoroff and Lusin type theorems in Vietoris hypertopology and also to point out several mathematical applications of these results. A Lebesgue type theorem for real- valued measurable functions with respect to $\mathcal{P}_0(X)$-valued fuzzy set multifunctions in the Vietoris hypertopology is also obtained. In the last two sections, some physical implications of the mathematical model of regularity are analyzed: regularization by sets of functions of ε-approximation type scale, which involves scale relativity type theories, regularization with known sets, which involves the trans-physics type theories and also simultaneous regularization with sets of functions of ε-approximation type scale and with known sets, which involves fractal string type theories.

8.2 Terminology, basic notions and results

For the beginning, we briefly recall some notions and notations involved in the Vietoris topology.

Let (X, τ) be a Hausdorff, linear topological space with the origin 0, T an abstract, nonvoid set and $\mathcal{C}$ a ring of subsets of T.

By $\mathcal{V}(0)$ ($\mathcal{U}(0)$, respectively) we denote the system (base—fundamental system, respectively) of neighbourhoods of 0.

The reader can refer to Chaps. 1 and 2 (see also Hu and Papageorgiou [18, Ch. 1], Precupanu et al. [40, Ch. 1], and [41, Ch. 8]):

$M^- = \{C \in \mathcal{P}_0(X);\ M \cap C \neq \emptyset\}$, $M^+ = \{C \in \mathcal{P}_0(X);\ C \subseteq M\}$,
$S_{UV} = \{D^+;\ D \in \tau\}$ and $S_{LV} = \{D^-;\ D \in \tau\}$.
Vietoris topology is denoted by $\widehat{\tau}_V$.

$\widehat{\tau}_V = \widehat{\tau}_V^+ \cup \widehat{\tau}_V^-$, where $\widehat{\tau}_V^+$ is *the upper Vietoris topology* and $\widehat{\tau}_V^-$ is *the lower Vietoris topology*.

$\widehat{\tau}_V^+$ has as a subbase the class S_{UV} and $\widehat{\tau}_V^-$ has as a subbase the class S_{LV}.

If $U, V \in \tau$, then the family of subsets $\mathcal{B}_{U,V} = \{M \in \mathcal{P}_0(X);\ M \subseteq U,\ M \cap V \neq \emptyset\}$ is a base for the topology $\widehat{\tau}_V$.

The family of subsets $\mathcal{B}_U = \{M \in \mathcal{P}_0(X);\ M \subseteq U\}$ (respectively, $\mathcal{B}_V = \{M \in \mathcal{P}_0(X);\ M \cap V \neq \emptyset\}$) is a base for $\widehat{\tau}_V^+$ (respectively, $\widehat{\tau}_V^-$).

Vietoris topology is an example of a hit-and-miss topology. Like some physical concepts, Vietoris topology, although it is composed of two independent parts of its, topologies $\widehat{\tau}_V^+$ and $\widehat{\tau}_V^-$, it becomes consistent when viewed as $\widehat{\tau}_V^+ \cup \widehat{\tau}_V^-$.

For instance, in physical terms, the non-differentiability of the curve motion of the physical object involves the simultaneous definition at any point of the curve, of two differentials (left and right). Since we cannot favour one of the two differentials, the only solution is to consider them simultaneously through a complex differential. Its application, multiplied by dt, where t is an affine parameter, to the field of space coordinates implies complex speed fields.

Definition 8.2.1 ([13]) A set multifunction $\mu : \mathcal{C} \to \mathcal{P}_0(X)$, with $\mu(\emptyset) = \{0\}$ is said to be a multisubmeasure if it is monotone and subadditive, i.e., $\mu(A \cup B) \subseteq \mu(A) + \mu(B)$, for every (disjoint) sets $A, B \in \mathcal{C}$.

Unless stated otherwise, all over this chapter we assume that (X, τ) is a Hausdorff, linear topological space and $\mu : \mathcal{C} \to \mathcal{P}_0(X)$ is a monotone set multifunction in the sense of Chap. 7.

We now recall from [13] the following notions and results concerning several types of continuity in the Vietoris topology, that will be used throughout this chapter.

Definition 8.2.2 ([13]) μ is said to be:

(i) $\widehat{\tau}_V^+$-decreasing convergent if for every decreasing sequence of sets $(A_n)_{n\in\mathbb{N}} \subset \mathcal{C}$, with $A_n \searrow A \in \mathcal{C}$, $\mu(A_n) \xrightarrow{\widehat{\tau}_V^+} \mu(A)$, i.e., for every $U \in \tau$, with $\mu(A) \subset U$, there exists $n_0(U) \in \mathbb{N}$ so that $\mu(A_n) \subset U$, for every $n \geq n_0$.

(ii) $\widehat{\tau}_V^-$-increasing convergent if for every increasing sequence of sets $(A_n)_{n\in\mathbb{N}} \subset \mathcal{C}$, with $A_n \nearrow A \in \mathcal{C}$, $\mu(A_n) \xrightarrow{\widehat{\tau}_V^-} \mu(A)$, i.e., for every $V \in \tau$, with $\mu(A) \cap V \neq \emptyset$, there exists $n_0(V) \in \mathbb{N}$ so that $\mu(A_n) \cap V \neq \emptyset$, for every $n \geq n_0$.

(iii) $\widehat{\tau}_V^+$-order continuous if for every decreasing sequence of sets $(A_n)_{n\in\mathbb{N}} \subset \mathcal{C}$, with $A_n \searrow \emptyset$, $\mu(A_n) \xrightarrow{\widehat{\tau}_V^+} \{0\}$, i.e., for every $U_0 \in \mathcal{V}(0)$, there exists $n_0(U_0) \in \mathbb{N}$ so that $\mu(A_n) \subset U_0$, for every $n \geq n_0$.

Proposition 8.2.3 ([13])

(i) If μ is $\widehat{\tau}_V^+$-decreasing convergent, then μ is $\widehat{\tau}_V^+$-order continuous.

(ii) If X is a metrizable linear topological space and μ is a $\widehat{\tau}_V^+$-order continuous multisubmeasure, then μ is $\widehat{\tau}_V^-$-increasing convergent and $\widehat{\tau}_V^+$-decreasing convergent.

(iii) If $\mathcal{C}$ is finite, then any set multifunction with $\mu(\emptyset) = \{0\}$ is $\widehat{\tau}_V^+$-decreasing convergent, $\widehat{\tau}_V^-$-increasing convergent and $\widehat{\tau}_V^+$-order continuous.

Lemma 8.2.4 ([13]) *If μ is $\widehat{\tau}_V^-$-increasing convergent, $U_0 \in \mathcal{V}(0)$, $(A_n)_{n\in\mathbb{N}} \subset \mathcal{C}$, with $A_n \nearrow A \in \mathcal{C}$ are arbitrary, fixed and if $\mu(A_n) \subset U_0$, for every $n \in \mathbb{N}$, then $\mu(A) \subset U_0$.*

We now establish the following lemma, that will be used in Sect. 8.3:

Lemma 8.2.5 *Suppose $\mathcal{C}$ is a σ-ring, μ is $\widehat{\tau_V}$-increasing convergent and $\widehat{\tau_V^+}$-decreasing convergent, $U_0 \in \mathcal{V}(0)$ is arbitrary and, for every $k \in \mathbb{N}^*$, $(B_n^{(k)})_n \subset \mathcal{C}$. Then:*

(i) If for every $k \in \mathbb{N}^*$, $B_n^{(k)} \underset{n\to\infty}{\searrow} \emptyset$, there exists $(n_k)_k$ so that $\mu(\overset{\infty}{\underset{k=1}{\cup}} B_{n_k}^{(k)}) \subset U_0$.

(ii) If, moreover, μ is a multisubmeasure and for every $k \in \mathbb{N}^*$, $B_n^{(k)} \underset{n\to\infty}{\searrow} C_k$, with $\mu(C_k) = \{0\}$, there exists $(n_k)_k$ so that $\mu(\overset{\infty}{\underset{k=1}{\cup}} B_{n_k}^{(k)}) \subset U_0$.

Proof

(i) Since $U_0 \in \mathcal{V}(0)$, there is $D_0 \in \tau$ so that $0 \in D_0 \subset U_0$ (evidently, $D_0 \in \mathcal{V}(0)$).

Because μ is $\widehat{\tau_V^+}$-decreasing convergent, by Proposition 8.2.3(i), it is also $\widehat{\tau_V^+}$-order continuous, so, since $B_n^{(1)} \underset{n\to\infty}{\searrow} \emptyset$, there exists $n_1 \in \mathbb{N}$ so that $\mu(B_{n_1}^{(1)}) \subset D_0 (\subset U_0)$. Then, since $B_{n_1}^{(1)} \cup B_n^{(2)} \underset{n\to\infty}{\searrow} B_{n_1}^{(1)}$ and μ is $\widehat{\tau_V^+}$-decreasing convergent, there exists $n_2 \in \mathbb{N}$ so that $n_2 > n_1$ and $\mu(B_{n_1}^{(1)} \cup B_{n_2}^{(2)}) \subset D_0 (\subset U_0)$.

Recurrently, there exists $(n_k)_k$ so that $\mu(\overset{s}{\underset{k=1}{\cup}} B_{n_k}^{(k)}) \subset D_0 (\subset U_0)$, for every $s \in \mathbb{N}^*$. Because μ is $\widehat{\tau_V}$-increasing convergent, by Lemma 8.2.4, we get that $\mu(\overset{\infty}{\underset{k=1}{\cup}} B_{n_k}^{(k)}) \subset U_0$.

(ii) For $U_0 \in \mathcal{V}(0)$, there exists $W_0 \in \mathcal{V}(0)$ so that $W_0 + W_0 \subset U_0$.

By (i) applied for W_0, there exists $(n_k)_k$ so that $\mu(\overset{\infty}{\underset{k=1}{\cup}} (B_{n_k}^{(k)} \backslash C_k)) \subset W_0$. Therefore,

$$\mu\left(\left(\overset{\infty}{\underset{k=1}{\cup}} B_{n_k}^{(k)}\right) \backslash \left(\overset{\infty}{\underset{k=1}{\cup}} C_k\right)\right) \subset \mu\left(\overset{\infty}{\underset{k=1}{\cup}} \left(B_{n_k}^{(k)} \backslash C_k\right)\right) \subset W_0.$$

On the other hand, because μ is a multisubmeasure, then, for every $s \in \mathbb{N}^*$, $\mu(\overset{s}{\underset{k=1}{\cup}} C_k) = \{0\}$, so, since μ is $\widehat{\tau_V}$-increasing convergent, by Lemma 8.2.4, $\mu(\overset{\infty}{\underset{k=1}{\cup}} C_k) \subset W_0$. Consequently,

$$\mu\left(\overset{\infty}{\underset{k=1}{\cup}} B_{n_k}^{(k)}\right) \subset \mu\left(\left(\overset{\infty}{\underset{k=1}{\cup}} B_{n_k}^{(k)}\right) \backslash \left(\overset{\infty}{\underset{k=1}{\cup}} C_k\right)\right) + \mu\left(\overset{\infty}{\underset{k=1}{\cup}} C_k\right) \subset$$
$$\subset W_0 + W_0 \subset U_0.$$

□

8.3 Convergences for real-valued measurable functions with respect to $\mathcal{P}_0(X)$-valued monotone set multifunctions in Vietoris topology

In the following, suppose $\mathcal{A}$ is a σ-algebra of subsets of T. Let be arbitrary $A \in \mathcal{A}$. By $A \cap \mathcal{A}$ we denote $\{B \in \mathcal{A}, B \subset A\}$.

We consider the class $\mathcal{M}$ of all $\mathcal{A}$-measurable real-valued functions on $(T, \mathcal{A}, \mu)$ (i.e., functions $f : T \to \mathbb{R}$ so that there exists a sequence $(f_n)_n$ of simple functions which converges to f on T). Let be arbitrary $f \in \mathcal{M}$ and $\{f_n\} \subset \mathcal{M}$.

Since $\mu(\emptyset) = \{0\}$ and μ is monotone, one can easily verify that it is natural to give only the following notions in the upper Vietoris topology $\widehat{\tau}_V^+$:

Definition 8.3.1 We say that $\{f_n\}$ *converges to* f *on* A :

(i) μ*-almost everywhere* (denoted by $f_n \xrightarrow[A]{a.e.} f$) if there exists a subset $B \in A \cap \mathcal{A}$ such that $\mu(B) = \{0\}$ and $\{f_n\}$ is pointwise convergent to f on $A \backslash B$.

(ii) *in* μ*-measure* (denoted by $f_n \xrightarrow[A]{\mu} f$) if for every $\varepsilon > 0$ and every $U \in \mathcal{V}(0)$, there exists $n_0(U, \varepsilon) \in \mathbb{N}$ so that for every $n \geq n_0$, $\mu(A_n(\varepsilon)) \subset U$, where $A_n(\varepsilon) = \{t \in A; |f_n(t) - f(t)| \geq \varepsilon\}$.

(iii) μ*-almost uniformly* (denoted by $f_n \xrightarrow[A]{a.u.} f$) if there exists a decreasing sequence $\{A_k\}_{k\in\mathbb{N}} \subset A \cap \mathcal{A}$, such that:

   (a) for every $U \in \mathcal{V}(0)$, there exists $k_0 = k_0(U) \in \mathbb{N}$ so that $\mu(A_{k_0}) \subset U$ and

   (b) for every fixed $k \in \mathbb{N}$, $\{f_n\}$ uniformly converges to f on $A \backslash A_k \big(f_n \xrightarrow[A\backslash A_k]{u} f\big)$.

Proposition 8.3.2 *If* $f_n \xrightarrow[A]{a.u.} f$ *and if in Definition 8.3.1(iii),* $A_k \searrow B$*, then* $\mu(B) = \{0\}$.

Proof By the hypothesis, for every $U \in \mathcal{U}(0)$, there exists $k_0(U) \in \mathbb{N}$ so that for every $k \geq k_0$, $\mu(A_k) \subset U$. Since $A_k \searrow B$ and μ is monotone, then $\mu(B) \subset \mu(A_{k_0}) \subset U$, whence $\{0\} \subset \mu(B) \subset \bigcap\limits_{U \in \mathcal{U}(0)} U = \{0\}$, so $\mu(B) = \{0\}$. □

Theorem 8.3.3

(i) If $f_n \xrightarrow[A]{a.u.} f$*, then* $f_n \xrightarrow[A]{a.e.} f$.

(ii) (Egoroff type) Conversely, if μ *is a* $\widehat{\tau}_V$*-increasing convergent and* $\widehat{\tau}_V^+$*-decreasing convergent multisubmeasure and if* $f_n \xrightarrow[A]{a.e.} f$*, then* $f_n \xrightarrow[A]{a.u.} f$.

Proof

(i) If $f_n \xrightarrow[A]{a.u.} f$, by Proposition 8.3.2, there exists a decreasing sequence $\{C_k\}_{k\in\mathbb{N}} \subset A \cap \mathcal{A}$ such that $C_k \searrow C$, $\mu(C) = \{0\}$ and for every fixed $k \in \mathbb{N}$, $\{f_n\}$ uniformly converges to f on $A \backslash C_k$.

For every $x \in A \backslash C$, there is $k_0 \in \mathbb{N}^*$ so that $x \in A \backslash C_{k_0}$ and, therefore, $f_n(x)$ converges to $f(x)$, whence $f_n \overset{a.e.}{\underset{A}{\to}} f$.

(ii) Let $f \in \mathcal{M}$, $\{f_n\}_{n\in\mathbb{N}} \subset \mathcal{M}$ be such that $f_n \overset{a.e.}{\underset{A}{\to}} f$. Let also $U \in \mathcal{V}(0)$ be arbitrary.

For every $m, n \in \mathbb{N}$, we define $A_n^{(m)} = \bigcup_{i=n}^{\infty} \{t \in A; |f_i(t) - f(t)| \geq \frac{1}{m}\}$.

We see that $A_n^{(m)} \supset A_{n+1}^{(m)}$, for every $m, n \in \mathbb{N}$ and $\mu(\bigcup_{m=1}^{\infty} \bigcap_{n=1}^{\infty} A_n^{(m)}) = \{0\}$, so $\mu(\bigcap_{n=1}^{\infty} A_n^{(m)}) = \{0\}$, for every $m \in \mathbb{N}$.

By Lemma 8.2.5(ii), there exists an increasing sequence $\{n_i\}_{i\in\mathbb{N}}$ of naturals such that $\mu(\bigcup_{i=1}^{\infty} A_{n_i}^i) \subset U$.

For every $k \in \mathbb{N}$, we denote $B_k = \bigcup_{i=k}^{\infty} A_{n_i}^i$. We observe that for every $k \in \mathbb{N}$, $B_k \supset B_{k+1}$, $U \supset \mu(B_k)$ and $\{f_n\}_{n\in\mathbb{N}}$ uniformly converges to f on every set $A \backslash B_k$. Consequently, $f_n \overset{a.u.}{\underset{A}{\longrightarrow}} f$. □

Corollary 8.3.4 *If $\mu : \mathcal{A} \to \mathcal{P}_0(X)$ is a $\widehat{\tau}_V$-increasing convergent and $\widehat{\tau}_V^+$-decreasing convergent multisubmeasure, then $f_n \overset{a.u.}{\underset{A}{\longrightarrow}} f$ iff $f_n \overset{a.e.}{\underset{A}{\longrightarrow}} f$.*

Theorem 8.3.5 (Lebesgue type) *Suppose X is a metrizable, linear topological space and $\mu : \mathcal{A} \to \mathcal{P}_0(X)$ is a multisubmeasure. Then $f_n \overset{a.e.}{\underset{A}{\longrightarrow}} f \Rightarrow f_n \overset{\mu}{\underset{A}{\longrightarrow}} f$ if and only if μ is $\widehat{\tau}_V^+$-decreasing convergent.*

Proof The set C of points $t \in A$ at which $\{f_n\}$ is pointwise convergent to f can be written as $C = \bigcap_{m=1}^{\infty} \bigcup_{n=1}^{\infty} \bigcap_{i=n}^{\infty} (A \backslash B_i(\frac{1}{m}))$, where $B_i(\frac{1}{m}) = \{t \in A; |f_i(t) - f(t)| \geq \frac{1}{m}\}$, for every $m, i \in \mathbb{N}^*$.

If for every $m, n \in \mathbb{N}^*$, we denote $A_n^{(m)} = \bigcup_{i=n}^{\infty} B_i(\frac{1}{m})$ and $A^{(m)} = \bigcap_{n=1}^{\infty} A_n^{(m)}$, then $A \backslash A^{(m)} = \bigcup_{n=1}^{\infty} (A \backslash A_n^{(m)}) = \bigcup_{n=1}^{\infty} \bigcap_{i=n}^{\infty} (A \backslash B_i(\frac{1}{m}))$ and $C = \bigcap_{m=1}^{\infty} \bigcup_{n=1}^{\infty} (A \backslash A_n^{(m)}) = \bigcap_{m=1}^{\infty} (A \backslash A^{(m)})$.

We observe that for every fixed $m \in \mathbb{N}^*$, $A_n^{(m)} \underset{n\to\infty}{\searrow} A^{(m)}$ and so, $A \backslash A_n^{(m)} \underset{n\to\infty}{\nearrow} A \backslash A^{(m)}$.

If there exists a set $B \in A \cap \mathcal{A}$ and $\{f_n\}$ is pointwise convergent to f on $A \backslash B$, then for every $m \in \mathbb{N}^*$,

$$A \backslash B \subset C \subset \bigcup_{n=1}^{\infty} \bigcap_{i=n}^{\infty} \left(A \backslash B_i \left(\frac{1}{m} \right) \right) = A \backslash A^{(m)} \subset A.$$

We also observe that $B_n(\frac{1}{m}) \subset A_n^{(m)}$, for every $m, n \in \mathbb{N}^*$.

The If part We prove that μ is $\widehat{\tau}_V^+$-decreasing convergent. By Proposition 8.2.3, this is equivalent to prove that μ is $\widehat{\tau}_V^+$-order continuous. So, we consider arbitrary $(A_n)_{n\in\mathbb{N}} \subset A \cap \mathcal{A}$, with $A_n \searrow \emptyset$ and arbitrary $U \in \mathcal{V}(0)$.

For every $n \in \mathbb{N}$, we define $f_n(t) = \begin{cases} 0, & \text{if } t \in A_n \\ 1, & \text{if } t \in A \backslash A_n. \end{cases}$

Evidently, $\{f_n\} \subset \mathcal{M}$ and $\{f_n\}$ is pointwise convergent to 1 on A, so $f_n \xrightarrow[A]{a.e.} 1$.

By the hypothesis, $f_n \xrightarrow[A]{\mu} f$, whence, for $\varepsilon = \frac{1}{2}$, there exists $n_0(U) \in \mathbb{N}^*$ so that, for every $n \geq n_0$, $\mu(A_n) = \mu(\{t \in A; |f_n(t) - 1| \geq \frac{1}{2}\}) \subset U$, that is, μ is $\widehat{\tau}_V^+$-order continuous, and so, equivalently, it is $\widehat{\tau}_V^+$-decreasing convergent.

The Only if part Suppose μ is $\widehat{\tau}_V^+$-decreasing convergent and $f_n \xrightarrow[A]{a.e.} f$.

Consequently, there exists $B \in A \cap \mathcal{A}$ such $\mu(B) = \{0\}$ and $\{f_n\}$ is pointwise convergent to f on $A \backslash B$. Therefore, for every $m \in \mathbb{N}^*$, $A^{(m)} \subset B$ and so, $\mu(A^{(m)}) = \{0\}$.

Let be arbitrary $U_0 \in \mathcal{V}(0)$. There is $D_0 \in \tau$ so that $0 \in D_0 \subset U_0$.

Since for every fixed $m \in \mathbb{N}^*$, $A_n^{(m)} \underset{n\to\infty}{\searrow} A^{(m)}$ and μ is $\widehat{\tau}_V^+$-decreasing convergent, then for every $m \in \mathbb{N}^*$, there exists $n_m(D_0) \in \mathbb{N}^*$ so that, for every $n \geq n_m$, $\mu(A_n^{(m)}) \subset D_0 \subset U_0$.

If $\varepsilon > 0$ is arbitrary, there exists $m_0(\varepsilon) \in \mathbb{N}^*$ so that $\frac{1}{m_0} < \varepsilon$, whence $B_n(\varepsilon) \subset B_n(\frac{1}{m_0})$. On the other hand, for every $m, n \in \mathbb{N}^*$, $B_n(\frac{1}{m}) \subset A_n^{(m)}$. Consequently, there exists $n_0(m_0, D_0) = n_0(U_0, \varepsilon) \in \mathbb{N}^*$ so that, for every $n \geq n_0$,

$$\mu(B_n(\varepsilon)) \subset \mu\left(B_n\left(\frac{1}{m_0}\right)\right) \subset \mu(A_n^{(m_0)}) \subset D_0 \subset U_0,$$

whence $f_n \xrightarrow[A]{\mu} f$. □

8.4 Set-valued Egoroff and Lusin type theorems and applications in the Vietoris topology

In what follows, without any other special assumptions, we suppose that (X, τ) is a Hausdorff, linear topological space with the origin 0, T is a locally compact Hausdorff space and $\mathcal{C}$ is a ring of subsets of T.

For the consistency of the following notions of regularity, one may suppose that $\mathcal{C}$ is, for instance, $\mathcal{B}_0$ (respectively, $\mathcal{B}_0'$) or $\mathcal{C}$ is $\mathcal{B}$ (respectively, $\mathcal{B}'$) (see Sect. 2.3).

$\mathcal{K}$ denotes the family of all compact subsets of T and $\mathcal{D}$, the family of all open subsets of T.

We now recall the following notions and results, that will be used throughout this section:

Remark 8.4.1

(i) Obviously, $\mathcal{B}_0 \subset \mathcal{B} \subset \mathcal{B}'$ and $\mathcal{B}_0 \subset \mathcal{B}'_0$.
Also, if T is metrizable or if it has a countable base, then any compact set $K \subset T$ is G_δ, so, in this case, $\mathcal{B}_0 = \mathcal{B}$ (Dinculeanu [9, Ch. III]) and, consequently, $\mathcal{B}'_0 = \mathcal{B}'$.

(ii) [13] For every $A \in \mathcal{C}$, there always exists $D \in \mathcal{D} \cap \mathcal{C}$ so that $A \subset D$.

(iii) [13] If $\mathcal{C}$ is $\mathcal{B}$ or $\mathcal{B}'$, then for every $A \in \mathcal{C}$, there always exist $K \in \mathcal{K} \cap \mathcal{C}$ and $D \in \mathcal{D} \cap \mathcal{C}$ so that $K \subset A \subset D$.

(iv) [39] If T is a compact, G_δ, Hausdorff topological space or a metrizable, compact space, then $\mathcal{B}'_0$ is a σ-algebra.

Definition 8.4.2 ([13])

(I) A set $A \in \mathcal{C}$ is said to be (with respect to μ):

  (i) $R_l'^{+}$*-regular* if for every $V_0 \in \mathcal{V}(0)$, there is $K = K_{V_0,A} \in \mathcal{K} \cap \mathcal{C}$ so that $K \subset A$ and $\mu(A \backslash K) \subset V_0$.

  (ii) R'^{+}*-regular* if for every $V_0 \in \mathcal{V}(0)$, there are $K = K_{V_0,A} \in \mathcal{K} \cap \mathcal{C}$ and $D = D_{V_0,A} \in \mathcal{D} \cap \mathcal{C}$ such that $K \subset A \subset D$ and $\mu(D \backslash K) \subset V_0$.

(II) μ is said to be $R_l'^{+}$*-regular* (R'^{+}*-regular,* respectively) if the same is every set $A \in \mathcal{C}$.

Example 8.4.3 Any set $K \in \mathcal{K} \cap \mathcal{C}$ is $R_l'^{+}$-regular.

In fact, the notions of regularity (given in the above definition in its abstract form) becomes consistent if it is considered for the case when $\mathcal{C}$ is $\mathcal{B}_0$, $\mathcal{B}$, $\mathcal{B}'$ or $\mathcal{B}'_0$. The local compacity of T is thus strongly needed.

Theorem 8.4.4 ([13]) *(Alexandroff type theorem) Suppose X is a metrizable, linear topological space. If μ is a $R_l'^{+}$-regular multisubmeasure, then μ is $\widehat{\tau}_V^{+}$-order continuous.*

We now prove a converse of Theorem 8.4.4 when $\mathcal{C}$ is the σ-ring $\mathcal{B}'_0$:

Theorem 8.4.5 *If T is a metrizable, locally compact space, X is a metrizable, linear topological space and if $\mu : \mathcal{B}'_0 \to \mathcal{P}_0(X)$ is a $\widehat{\tau}_V^{+}$-order continuous multisubmeasure, then μ is R'^{+}-regular on $\mathcal{B}'_0$.*

Proof Denote $\mathcal{A} = \{A \in \mathcal{B}'_0; A$ is R'^{+}-regular with respect to $\mu\}$.

We shall prove that $\mathcal{A} = \mathcal{B}'_0$. By the proof of Theorem 4.13 from [13], $\mathcal{A}$ is nonvoid, and, also, by Lemma 4.12 [13], $\mathcal{A}$ is a δ-ring which contains all G_δ, compact sets of T.

We now prove that $\mathcal{A}$ is a σ-ring. For this, it is sufficient to establish that for every increasing sequence of sets $(A_n)_{n\in\mathbb{N}} \subset \mathcal{A}$, with $A_n \nearrow A$, we have $A \in \mathcal{A}$.

Since X is a metrizable linear topological space, by Precupanu [37] there exists a fundamental system of neighbourhoods of the origin, $\mathcal{U}(0) = \{V_n\}_{n\in\mathbb{N}}$ so that, for every arbitrary fixed $n \in \mathbb{N}$,

$$V_{n+1} + V_{n+2} + \cdots + V_{n+p} \subset V_n, \text{ for every } p \in \mathbb{N}^*.$$

Consider arbitrary $V \in \mathcal{V}(0)$. There exists $\widetilde{V} \in \mathcal{U}(0)$ so that $\widetilde{V} \subset V$. Obviously, there is $i_0 \in \mathbb{N}$ so that $\widetilde{V} = V_{i_0}$ and for every $p \in \mathbb{N}^*$,

$$V_{i_0+1} + V_{i_0+2} + \cdots + V_{i_0+p} \subset \widetilde{V} \subset V.$$

Since for every $r \in \mathbb{N}$, A_r is R'^{+}-regular, there exist $K_r \in \mathcal{K} \cap \mathcal{B}'_0$ and $D_r \in \mathcal{D} \cap \mathcal{B}'_0$ so that $K_r \subset A_r \subset D_r$ and $\mu(D_r \backslash K_r) \subset V_{i_0+r+1}$.

Denote $C_r = \overset{r}{\underset{i=1}{\cup}} K_i$, $E_r = \overset{r}{\underset{i=1}{\cup}} D_i$ and $D = \overset{\infty}{\underset{i=1}{\cup}} E_r$. Evidently, for every $r \in \mathbb{N}^*$, $C_r \in \mathcal{K} \cap \mathcal{B}'_0$ and $E_r, D \in \mathcal{D} \cap \mathcal{B}'_0$.

Since $E_r \nearrow D$ and μ is $\widehat{\tau}_V^+$-order continuous, there exists $r_0 \in \mathbb{N}^*$ so that $\mu(D \backslash E_{r_0}) \subset V_{i_0+1}$.

We observe that $C_{r_0} \subset E_{r_0} \subset D$, $C_{r_0} \subset A \subset D$ and $E_{r_0} \backslash C_{r_0} = \big(\overset{r_0}{\underset{i=1}{\cup}} D_i\big) \backslash \big(\overset{r_0}{\underset{i=1}{\cup}} K_i\big) \subset \overset{r_0}{\underset{i=1}{\cup}} (D_i \backslash K_i)$, so

$$\mu(E_{r_0} \backslash C_{r_0}) \subset \mu(D_1 \backslash K_1) + \cdots + \mu(D_{r_0} \backslash K_{r_0}) \subset V_{i_0+2} + \cdots + V_{i_0+r_0+1}.$$

Then

$$\mu(D \backslash C_{r_0}) = \mu((D \backslash E_{r_0}) \cup (E_{r_0} \backslash C_{r_0})) \subset$$
$$\subset V_{i_0+1} + V_{i_0+2} + \cdots + V_{i_0+r_0+1} \subset \widetilde{V} \subset V,$$

so, A is R'^{+}-regular.

Consequently, $\mathcal{A} \subset \mathcal{B}'_0$ is a σ-ring which contains all G_δ, compact sets of T, so $\mathcal{A} = \mathcal{B}'_0$. This means μ is R'^{+}-regular on $\mathcal{B}'_0$. □

We are now able to establish our set-valued variants of Egoroff and Lusin type theorems for set multifunctions in the Vietoris topology:

Theorem 8.4.6 (Egoroff type theorem) *Suppose $\mathcal{C}$ is a σ-algebra of subsets of a locally compact Hausdorff space T and $\mu : \mathcal{C} \to \mathcal{P}_0(X)$ is a $\widehat{\tau}_V^+$-decreasing convergent, $\widehat{\tau}_V^-$-increasing convergent and $R_l'^{+}$-regular multisubmeasure. If $(f_n) \subset \mathcal{M}$, $f \in \mathcal{M}$ are so that (f_n) converges to f μ-almost everywhere on $A \in \mathcal{C}$, then for every $U \in \mathcal{V}(0)$, there exists $K = K_U \in \mathcal{K} \cap \mathcal{C}$ so that $\mu(A \backslash K) \subset U$ and (f_n) uniformly converges to f on K.*

Proof Generally, if $\mu(M) = \{0\}$ and $\mu(N) \subset U$ (where $M, N \in \mathcal{C}$ are disjoint), then $\mu(M \cup N) \subset U$, so, we can reduce our considerations to the case when (f_n) converges to f everywhere on T.

Consider arbitrary $U \in \mathcal{V}(0)$. There exists $W \in \mathcal{V}(0)$ so that $W + W \subset U$.

If $\widetilde{A}$ denotes the set of points $t \in T$ so that $f_n(t) \to f(t)$, we observe that $\widetilde{A} = \bigcap_{k=1}^{\infty} \bigcup_{n=1}^{\infty} A_n^{(k)}$, where for every $k \in \mathbb{N}^*$, $A_n^{(k)} = \bigcap_{j=n}^{\infty} \{t \in T; |f_j(t) - f(t)| < \frac{1}{k}\}$.

Since (f_n) converges to f everywhere on T, then for every $k \in \mathbb{N}^*$, $A_n^{(k)} \underset{n\to\infty}{\nearrow} T$, whence $T \setminus A_n^{(k)} \underset{n\to\infty}{\searrow} \emptyset$.

By Lemma 8.2.5(i), for $W \in \mathcal{V}(0)$, there exists $(n_k)_k$ so that $\mu(\bigcup_{k=1}^{\infty} (T \setminus A_{n_k}^{(k)})) \subset W$.

If we denote $B = \bigcap_{k=1}^{\infty} A_{n_k}^{(k)}$, then $\mu(T \setminus B) = \mu(\bigcup_{k=1}^{\infty} (T \setminus A_{n_k}^{(k)})) \subset W$. We also observe that (f_n) uniformly converges to f on B.

On the other hand, since μ is $R_l^{\prime+}$-regular, for B there exists $K \in \mathcal{K} \cap \mathcal{C}$ so that $K \subset B$ and $\mu(B \setminus K) \subset W$.

Obviously, (f_n) uniformly converges to f on K and, because μ is a multisubmeasure, we have

$$\mu(T \setminus K) \subset \mu(T \setminus B) + \mu(B \setminus K) \subset W + W \subset U.$$

□

By Example 8.4.3, Proposition 8.2.3(ii), Remark 8.4.1(iv) and Theorem 8.4.5, we get:

Theorem 8.4.7 (Egoroff type theorem) *Suppose T is a metrizable, compact space, X is a metrizable, linear topological space and $\mu : \mathcal{B}_0' \to \mathcal{P}_0(X)$ is a $R_l^{\prime+}$-regular multisubmeasure. If $(f_n) \subset \mathcal{M}$, $f \in \mathcal{M}$ are so that (f_n) converges to f μ-almost everywhere on T, then for every $U \in \mathcal{V}(0)$, there exists $K = K_U \in \mathcal{K} \cap \mathcal{B}_0'$ so that $\mu(T \setminus K) \subset U$ and (f_n) uniformly converges to f on K.*

We now state our two main results:

Theorem 8.4.8 (Lusin type theorem) *Suppose $\mathcal{C}$ is a σ-algebra of subsets of a normal, locally compact Hausdorff space T, X is a metrizable, linear topological space and $\mu : \mathcal{C} \to \mathcal{P}_0(X)$ is a $\widehat{\tau}_V^+$-decreasing convergent, $\widehat{\tau}_V^-$-increasing convergent and $R_l^{\prime+}$-regular multisubmeasure. If $f \in \mathcal{M}$, then for every $U \in \mathcal{V}(0)$, there exists $K = K_U \in \mathcal{K} \cap \mathcal{C}$ so that $\mu(T \setminus K) \subset U$ and f is continuous on K.*

Proof Let be arbitrary $U \in \mathcal{V}(0)$. We prove the theorem stepwise:

Step 1 Suppose f is a simple function, i.e., for every $t \in T$, $f(t) = \sum_{i=1}^{p} a_i \aleph_{A_i}$, where $(A_i)_{i=\overline{1,p}}$ is a partition of T.

For $U \in \mathcal{V}(0)$, there exists $V_p \in \mathcal{V}(0)$ so that $\underbrace{V_p + V_p + \cdots + V_p}_{p} \subset U$.

Since μ is $R_l'^{+}$-regular, then for every $i = \overline{1, p}$, there exists $K_i \in \mathcal{K} \cap \mathcal{C}$ so that $K_i \subset A_i$ and $\mu(A_i \backslash K_i) \subset V_p$. If we denote $K = \bigcup_{i=1}^{p} K_i$, then $K \in \mathcal{K} \cap \mathcal{C}$ and, since μ is a multisubmeasure,

$$\mu(T \backslash K) \subset \mu\left(\bigcup_{i=1}^{p} (A_i \backslash K_i)\right) \subset \underbrace{V_p + V_p + \cdots + V_p}_{p} \subset U.$$

Moreover, since T is normal, then f is continuous on K.

Step 2 Suppose $f \in \mathcal{M}$ *is arbitrary.* Then f is the limit of an increasing sequence (f_m) of simple functions.

For $U \in \mathcal{V}(0)$, there exists $W \in \mathcal{V}(0)$ so that $W + W \subset U$. For W there exists $W' \in \mathcal{U}(0)$ so that $W' \subset W$.

On the other hand, since X is a metrizable, linear topological space, by Precupanu [37] there exists a countable fundamental system of vicinities of the origin, $\mathcal{U}(0) = \{V_n\}_{n \in \mathbb{N}}$ so that, for every arbitrary fixed $n \in \mathbb{N}$,

$$(*)\ V_{n+1} + V_{n+2} + \cdots + V_{n+p} \subset V_n, \text{ for every } p \in \mathbb{N}^*.$$

Evidently, there exists $n_0 \in \mathbb{N}$ so that $W' = V_{n_0}$, so, $V_{n_0+1} + V_{n_0+2} + \cdots + V_{n_0+p} \subset V_{n_0} = W' \subset W$, for every $p \in \mathbb{N}^*$.

By Step 1, there exists a sequence $(K_m) \subset \mathcal{K} \cap \mathcal{C}$ so that, for every $m \in \mathbb{N}^*$, $\mu(T \backslash K_m) \subset V_{n_0+m}$ and f_m is continuous on K_m.

Consequently, since μ is a multisubmeasure, for every $s \in \mathbb{N}^*$,

$$\mu\left(\bigcup_{m=1}^{s} (T \backslash K_m)\right) \subset \mu(T \backslash K_1) + \mu(T \backslash K_2) + \cdots + \mu(T \backslash K_s) \subset$$
$$\subset V_{n_0+1} + V_{n_0+2} + \cdots + V_{n_0+s} \subset V_{n_0} = W.$$

By Lemma 8.2.4, we get that $\mu(\bigcup_{m=1}^{\infty} (T \backslash K_m)) \subset W$, so, if we denote $\widetilde{K} = \bigcap_{m=1}^{\infty} K_m$, then

$$\mu(T \backslash \widetilde{K}) = \mu\left(\bigcup_{m=1}^{\infty} (T \backslash K_m)\right) \subset W.$$

On the other hand, by Theorem 8.4.5, there exists $K_0 \in \mathcal{K} \cap \mathcal{C}$ so that $\mu(T \backslash K_0) \subset W$ and $f_m \xrightarrow[K_0]{u} f$. Denote $K = \widetilde{K} \cap K_0 \in \mathcal{K} \cap \mathcal{C}$. Then $f_m \xrightarrow[K]{u} f$ and

$$\mu(T \backslash K) \subset \mu(T \backslash K_0) + \mu(T \backslash \widetilde{K}) \subset W + W \subset U,$$

so, the proof finishes. □

Remark 8.4.9 Let us observe that, in the above theorem, although in some particular cases regularity is equivalent to order continuity/decreasing convergence, in this situation regularity is needed since, on the one hand, $\mathcal{C}$ is here an arbitrary σ-algebra and, on the other hand, the existence of the compact sets $K_i \in \mathcal{K} \cap \mathcal{C}, i = \overline{1, p}$, cannot be otherwise assured.

By Remark 8.4.1(iv) and Theorem 8.4.7, we get:

Corollary 8.4.10 (Lusin type theorem) *Suppose T is a metrizable, compact space, X is a metrizable, linear topological space and $\mu : \mathcal{B}'_0 \to \mathcal{P}_0(X)$ is a $R_l'^{+}$-regular multisubmeasure. If $f \in \mathcal{M}$, then for every $U \in \mathcal{V}(0)$, there exists $K = K_U \in \mathcal{K} \cap \mathcal{B}'_0$ so that $\mu(T \backslash K) \subset U$ and f is continuous on K.*

We now present several applications of the set-valued Lusin theorems in the Vietoris topology. In what follows, we suppose that the conditions of Remark 8.4.9 are fulfilled.

Corollary 8.4.11 *If $f \in \mathcal{M}$, there exists a sequence $(f_n)_n$ of continuous functions on T so that $f_n \underset{T}{\overset{\mu}{\to}} f$. Moreover, if $|f| \le M$, then $|f_n| \le M$, for every $n \in \mathbb{N}$.*

Proof Evidently, $\mathcal{U}(0) = \{V_n\}_{n \ge 1}$, where $\bigcap_{n=1}^{\infty} V_n = \{0\}$ and $V_i \supset V_{i+1}$, for every $i \in \mathbb{N}^*$.

For every $U \in \mathcal{V}(0)$, there exists $W \in \mathcal{U}(0)$, with $W = V_{i_0} \subset U$.

By Corollary 8.4.9, for every $n \ge 1$, there is a compact set $K_n \subset T$ so that $\mu(T \backslash K_n) \subset V_n$ and f is continuous on K_n. By Tietze's extension theorem, for every $n \ge 1$, there is $f_n : T \to \mathbb{R}$ so that f_n is continuous on T, $f_n(t) = f(t)$, for every $t \in K_n$ and if $|f| \le M$, then $|f_n| \le M$.

Consequently, for every $\varepsilon > 0$ and every $U \in \mathcal{V}(0)$, there exists $i_0 = i_0(U)(= i_0(U, \varepsilon)) \in \mathbb{N}$ so that for every $n \ge i_0$,

$$\mu(\{t \in T; |f_n(t) - f(t)| \ge \varepsilon\}) \subset \mu(T \backslash K_n) \subset V_n \subset V_{i_0} = W \subset U,$$

whence $f_n \underset{T}{\overset{\mu}{\to}} f$. □

Corollary 8.4.12 *If $f \in \mathcal{M}$, there exists a sequence $(P_n)_n$ of polynomials on $[a, b]$ so that $P_n \underset{[a,b]}{\overset{\mu}{\to}} f$. Moreover, if $|f| \le M$, then $|P_n| \le M + 1$, for every $n \in \mathbb{N}$.*

Proof By Corollary 8.4.10 on the reduced space $([a, b], [a, b] \cap \mathcal{B}'_0, \mu)$, there exists a sequence $(f_n)_n$ of continuous functions on $[a, b]$, so that $f_n \underset{[a,b]}{\overset{\mu}{\to}} f$.

Consequently, there exists a subsequence $(f_{n_k})_k$ of $(f_n)_n$ so that, for every $k \ge 1$, $\mu\left(\left\{t \in T; |f_{n_k}(t) - f(t)| \ge \frac{1}{2k}\right\}\right) \subset V_k$.

Applying Weierstrass' theorem for every $f_{n_k}, k \ge 1$, on $[a, b]$, there exists a polynomial P_k on $[a, b]$ so that $|f_{n_k}(t) - P_k(t)| < \frac{1}{2k}$, whence $\mu\left(\left\{t \in T; |f_{n_k}(t) - P_k(t)| \ge \frac{1}{2k}\right\}\right) = \mu(\emptyset) = \{0\}$.

Therefore, for every $k \geq 1$,

$$\mu\left(\left\{t \in T; |f(t) - P_k(t)| \geq \frac{1}{k}\right\}\right) \subset \mu\left(\left\{t \in T; |f_{n_k}(t) - f(t)| \geq \frac{1}{2k}\right\}\right)$$
$$+\mu\left(\left\{t \in T; |f_{n_k}(t) - P_k(t)| \geq \frac{1}{2k}\right\}\right)$$
$$= \mu\left(\left\{t \in T; |f_{n_k}(t) - f(t)| \geq \frac{1}{2k}\right\}\right) \subset V_k.$$

We shall prove that $P_n \underset{[a,b]}{\overset{\mu}{\rightarrow}} f$. Indeed, for every $U \in \mathcal{V}(0)$, there exists $W \in \mathcal{U}(0)$ so that $W \subset U$. Evidently, $W = V_{i'_0}$, where $i'_0 = i'_0(U)$ and for every $i \geq i'_0$,

$$\mu\left(\left\{t \in T; |f(t) - P_i(t)| \geq \frac{1}{i}\right\}\right) \subset V_i \subset V_{i'_0} = W \subset U.$$

On the other hand, for every $\varepsilon > 0$, there exists $i''_0 = i''_0(\varepsilon)$ so that $\frac{1}{i} < \varepsilon$, for every $i \geq i''_0$.

Let be $i_0 = i_0(U, \varepsilon) = \max\{i'_0, i\}$. Then, for every $i \geq i_0$,

$$\mu(\{t \in T; |f(t) - P_i(t)| \geq \varepsilon\}) \subset \mu\left(\left\{t \in T; |f(t) - P_i(t)| \geq \frac{1}{i}\right\}\right) \subset U,$$

which says that $P_n \underset{[a,b]}{\overset{\mu}{\rightarrow}} f$.

Moreover, if $|f| \leq M$, then, according to Corollary 8.4.10, $|f_{n_k}| \leq M$, for every $k \geq 1$.

Since for every $t \in [a, b]$ and every $k \geq 1$, we have $|P_k(t) - f_{n_k}(t)| < \frac{1}{2k}$, then $|P_k(t)| \leq M + 1$. □

8.5 Regularization by sets of functions of ε-approximation type scale

Let us consider a fractal function $f(x)$ (see [33, 34]) with $x \in [a, b]$, for instance, one of the trajectory's equation. We consider now the sequence of the values of the variable x :

$$x_a = x_0, x_1 = x_0 + \varepsilon, \ldots, x_k = x_0 + k\varepsilon, \ldots, x_n = x_0 + n\varepsilon = x_b \tag{8.1}$$

We shall denote by $f(x, \varepsilon)$, the fractured (broken) line connecting the points

$$f(x_0), \ldots, f(x_k), \ldots, f(x_n).$$

The broken line will be considered as an approximation which is different from the one used before. We shall say that $f(x, \varepsilon)$ is an *ε-approximation scale.*

Let us consider now the $\overline{\varepsilon}$-approximation scale $f(x, \overline{\varepsilon})$ of the same function. Since $f(x)$ is similar almost everywhere, if ε and $\overline{\varepsilon}$ are small enough, then the two approximations $f(x, \varepsilon)$ and $f(x, \overline{\varepsilon})$ must lead to the same results when we study a fractal phenomenon by approximation. If we compare the two cases, then to an infinitesimal increase $d\varepsilon$ of ε, it corresponds to an increase $d\overline{\varepsilon}$ of $\overline{\varepsilon}$, if the scale is dilated. But in this case we have $\frac{d\varepsilon}{\varepsilon} = \frac{d\overline{\varepsilon}}{\overline{\varepsilon}}$, that is,

$$\frac{d\varepsilon}{\varepsilon} = d\rho \tag{8.2}$$

is the ratio of the scale $\varepsilon + d\varepsilon$ and $d\varepsilon$ must be preserved. We can then consider the infinitesimal transformation of the scale as

$$\varepsilon' = \varepsilon + d\varepsilon = \varepsilon + \varepsilon d\rho. \tag{8.3}$$

By such transformation, in the case of the function $f(x, \varepsilon)$, it results:

$$f(x, \varepsilon') = f(x, \varepsilon + \varepsilon d\rho), \tag{8.4}$$

respectively, if we stop after the first approximation,

$$f(x, \varepsilon') = f(x, \varepsilon) + \frac{\partial f}{\partial \varepsilon}(\varepsilon' - \varepsilon), \tag{8.5}$$

that is,

$$f(x, \varepsilon') = f(x, \varepsilon) + \frac{\partial f}{\partial \varepsilon}\varepsilon d\rho. \tag{8.6}$$

Let us also observe that, for arbitrary, but fixed ε_0,

$$\frac{\partial \ln \frac{\varepsilon}{\varepsilon_0}}{\partial \varepsilon} = \frac{\partial(\ln \varepsilon - \ln \varepsilon_0)}{\partial \varepsilon} = \frac{1}{\varepsilon}, \tag{8.7}$$

so, (8.5) becomes

$$f(x, \varepsilon') = f(x, \varepsilon) + \frac{\partial f(x, \varepsilon)}{\partial \ln \frac{\varepsilon}{\varepsilon_0}} d\rho. \tag{8.8}$$

Finally, we have

$$f(x, \varepsilon') = \left(1 + \frac{\partial}{\partial \ln \frac{\varepsilon}{\varepsilon_0}} d\rho\right) f(x, \varepsilon). \tag{8.9}$$

The operator

$$\widetilde{D} = \frac{\partial}{\partial \ln \frac{\varepsilon}{\varepsilon_0}} \tag{8.10}$$

is called the *dilation operator*.

The relation (8.10) shows us that the intrinsic variable of the resolution is not ε, but $\ln \frac{\varepsilon}{\varepsilon_0}$.

8.6 Physical implications

Simultaneous invariance with respect to both space-time coordinates and the resolution scale induces general relativity theory of type scale [30, 31]. These theories are more general than Einstein's theory of general relativity, being invariant with respect to the generalized Poincaré group (standard Poincaré group and dilatation group) [17, 30, 31, 36].

Basically, we discuss various physical theories built on varieties of space-time fractal type. They all turn out to be reducible to one of the following classes:

(i) Scale relativity theory [33, 34] and its possible extensions [1]. It is considered that microparticles' motion occurs on non-differentiable continuous curves. In such context, regularization works using sets of functions of ε-approximation type scale.

(ii) Transition in which to each point of the motion trajectory, a transfinite set is assigned (in particular, a Cantor type set—see El Naschie et al. $\varepsilon^{(\infty)}$ model of space-time [32]), in order to mimic the continuous (the trans-physics). In such context, the regularization of "vague" sets by known sets works.

(iii) Fractal string theories containing simultaneously both relativity and trans-physic [17, 36].

The reduction of the complex dimensions to their real part is equivalent to the theories of relativity type scale, while reducing them to their imaginary part of their complex dimensions generates the trans-physics. In such context, the simultaneous regularization by sets of functions of ε-approximation type scale and also by "known" sets works. The "reduction" of the complex dimensions to their real part requires the regularization by sets of functions of ε-approximation type scale, while the "reduction" to their imaginary part requires regularization with "known" sets.

8.7 Conclusions

In this chapter, some mathematical applications of regularity in Vietoris topology are established and several physical implications of the mathematical model of regularity are presented, which allows a classification of the physical models.

References

1. Agop, M., Niculescu, O., Timofte, A., Bibire, L., Ghenadi, A.S., Nicuta, A., Nejneru, C., Munceleanu, G.V.: Non-differentiable mechanical model and its implications. Int. J. Theor. Phys. **49**(7), 1489–1506 (2010)
2. Andres, J., Fiser, J.: Metric and topological multivalued fractals. Int. J. Bifur. Chaos Appl. Sci. Eng. **14**(4), 1277–1289 (2004)
3. Andres, J., Rypka, M.: Multivalued fractals and hyperfractals. Int. J. Bifur. Chaos Appl. Sci. Eng. **22**(1), 1250009, 27 pp. (2012)
4. Apreutesei, G.: Families of subsets and the coincidence of hypertopologies. Ann. Alexandru Ioan Cuza Univ. Math. **XLIX**, 1–18 (2003)
5. Averna, D.: Lusin type theorems for multifunctions. Scorza Dragoni's property and Carathéodory selections. Boll. U.M.I. **7**(8-A), 193–201 (1994)
6. Banakh, T., Novosad, N.: Micro and macro fractals generated by multi-valued dynamical systems. arXiv: 1304.7529v1 [math.GN], 28 April 2013
7. Beer, G.: Topologies on Closed and Closed Convex Sets. Kluwer, Dordrecht (1993)
8. Brown, S.: Memory and mathesis: for a topological approach to psychology. Theory Cult. Soc. **29**(4–5), 137–164 (2012)
9. Dinculeanu, N.: Measure Theory and Real Functions (in Romanian). Ed. Did. şi Ped., Bucureşti (1964)
10. Fu, H., Xing, Z.: Mixing properties of set-valued maps on hyperspaces via Furstenberg families. Chaos, Solitons Fractals **45**(4), 439–443 (2012)
11. Gavriluţ, A.: A Lusin type theorem for regular monotone uniformly autocontinuous set multifunctions. Fuzzy Sets Syst. **161**, 2909–2918 (2010)
12. Gavriluţ, A.: Fuzzy Gould integrability on atoms. Iran. J. Fuzzy Syst. **8**(3), 113–124 (2011)
13. Gavriluţ, A.: Continuity properties and Alexandroff theorem in Vietoris topology. Fuzzy Sets Syst. **194**, 76–89 (2012)
14. Gavriluţ, A.: Alexandroff theorem in Hausdorff topology for null-null-additive set multifunctions. Ann. Alexandru Ioan Cuza Univ. Math. **LIX**(2), 237–251 (2013)
15. Gómez-Rueda, J.L., Illanes, A., Méndez, H.: Dynamic properties for the induced maps in the symmetric products. Chaos, Solitons Fractals **45**(9–10), 1180–1187 (2012)
16. Guo, C., Zhang, D.: On the set-valued fuzzy measures. Inform. Sci. **160**, 13–25 (2004)
17. Hawking, S., Penrose, R.: The Nature of Space Time. Princeton University Press, Princeton (1996)
18. Hu, S., Papageorgiou, N.S.: Handbook of Multivalued Analysis, vol. I. Kluwer, Dordrecht (1997)
19. Jiang, Q., Suzuki, H.: Fuzzy measures on metric spaces. Fuzzy Sets Syst. **83**, 99–106 (1996)
20. Kawabe, J.: Regularity and Lusin's theorem for Riesz space-valued fuzzy measures. Fuzzy Sets Syst. **158**, 895–903 (2007)
21. Kunze, H., La Torre, D., Mendivil, F., Vrscay, E.R.: Fractal Based Methods in Analysis. Springer, New York (2012)
22. Lewin, K., Heider, G.M., Heider, F.: Principles of Topological Psychology. McGraw-Hill, New York (1936)
23. Li, R.: A note on stronger forms of sensitivity for dynamical systems. Chaos, Solitons Fractals **45**(6), 753–758 (2012)
24. Li, J., Yasuda, M.: Lusin's theorem on fuzzy measure spaces. Fuzzy Sets Syst. **146**, 121–133 (2004)
25. Li J., Gao S., Chen B.: Approximation of fuzzy neural networks to fuzzy-valued measurable function. In: Cao B., Wang G., Chen S., Guo S. (eds.) Quantitative Logic and Soft Computing 2010. Advances in Intelligent and Soft Computing, vol. 82. Springer, Berlin (2010)
26. Liu, L., Wang, Y., Wei, G.: Topological entropy of continuous functions on topological spaces. Chaos, Solitons Fractals **39**(1), 417–427 (2009)

27. di Lorenzo, P., di Maio, G.: The Hausdorff metric in the melody space: a new approach to melodic similarity. In: 9th the International Conference on Music Perception and Cognition, Alma Mater Studiorum University of Bologna, August 22–26, 2006
28. Lu, Y., Tan, C.L., Huang, W., Fan, L.: An approach to word image matching based on weighted Hausdorff distance. In: Conference: Document Analysis and Recognition, 2001. Proceedings., February 2001. https://doi.org/10.1109/ICDAR.2001.953920
29. Ma, X., Hou, B., Liao, G.: Chaos in hyperspace system. Chaos, Solitons Fractals **40**(2), 653–660 (2009)
30. El-Nabulsi, A.R.: New astrophysical aspects from Yukawa fractional potential correction to the gravitational potential in D dimensions. Indian J. Phys. **86**, 763–768 (2012)
31. El-Nabulsi, A.R.: Fractional derivatives generalization of Einstein's field equations. Indian J. Phys. **87**, 195–200 (2013)
32. El Naschie, M.S., Rösler, O.E., Prigogine, I. (eds.): Quantum Mechanics, Diffusion and Chaotic Fractals. Elsevier, Oxford (1995)
33. Nottale, L.: Fractal Space-Time and Microphysics: Towards Theory of Scale Relativity. World Scientific, Singapore (1993)
34. Nottale, L.: Scale Relativity and Fractal Space-Time. A New Approach to Unifying Relativity and Quantum Mechanics. Imperial College Press, London (2011)
35. Pap, E.: Null-additive Set Functions. Kluwer, Dordrecht (1995)
36. Penrose, R.: The Road to Reality: A Complete Guide to the Laws of the Universe. Jonathan Cape, London (2004)
37. Precupanu, T.: Linear Topological Spaces and Elements of Convex Analysis (in Romanian). Ed. Acad. Romania (1992)
38. Precupanu, A., Gavriluţ, A.: A set-valued Egoroff type theorem. Fuzzy Sets Syst. **175**, 87–95 (2011)
39. Precupanu, A., Gavriluţ, A.: A set-valued Lusin type theorem. Fuzzy Sets Syst. **204**, 106–116 (2012)
40. Precupanu., A., Gavriluţ., A.: Setvalued Lusin type theorem for null-null-additive set multifunctions. Fuzzy Sets Syst. **204**, 106–116 (2012)
41. Precupanu, A., Precupanu, T., Turinici, M., Apreutesei Dumitriu, N., Stamate, C., Satco, B.R., Văideanu, C., Apreutesei, G., Rusu, D., Gavriluţ, A.C., Apetrii, M.: Modern Directions in Multivalued Analysis and Optimization Theory. Venus Publishing House, Iaşi (2006) (in Romanian)
42. Sharma, P., Nagar, A.: Topological dynamics on hyperspaces. Appl. General Topology **11**(1), 1–19 (2010)
43. Song, J., Li, J.: Regularity of null-additive fuzzy measure on metric spaces. Int. J. Gen. Syst. **32**, 271–279 (2003)
44. Wang, Y., Wei, G., Campbell, W.H., Bourquin, S.: A framework of induced hyperspace dynamical systems equipped with the hit-or-miss topology. Chaos, Solitons Fractals **41**(4), 1708–1717 (2009)
45. Wicks, K.R.: Fractals and Hyperspaces. Springer, Berlin/Heidelberg (1991)

Chapter 9
Atomicity via regularity for non-additive set multifunctions

In this chapter, an approach of atomicity problems is presented by means of regularity. Characterizations and physical interpretations of atoms and non-atomicity for set multifunctions taking values in the family of all nonempty subsets of a topological space are given.

In recent literature, by many important theoretical and practical necessities, classical problems from measure theory were transferred to the non-additive (set-valued) (fuzzy) measures framework (Agahi et al. [1], Mesiar et al. [35, 36], Pap et al. [43], Gagolewski and Mesiar [10], Maghsoudi [33], etc.). This general setting requires the use of a non-additive set multifunction taking values in the family of all nonempty subsets of a topological space, family which is usually endowed with a hypertopology (like Hausdorff-Pompeiu, Vietoris, Wijsman or many others known in literature) (Apreutesei [2], Beer [4], Hu [23]).

In this context, regularity is known as an important continuity property with respect to such different (hyper)topologies, but, at the same time, it can be interpreted as an approximation property. Precisely, in this way, we can approximate no matter how well different "unknown" sets by other sets for which we have more information. Usually, from a mathematical perspective, this approximation is done from the left by closed sets (also called *inner regularity*), or more restrictive, by compact sets and/or from the right by open sets (also *called inner regularity/outer regularity/regularity*). As a mathematical direct application of regularity, the classical Lusin's theorem concerning the existence of continuous restrictions of measurable functions is very important and useful for discussing different kinds of approximation of measurable functions defined on special topological spaces and for numerous applications in the study of convergence of sequences of Sugeno and Choquet integrable functions (Croitoru [8], Dinculeanu [9], Gavriluţ and Agop [16], Gavriluţ [17–19], Gong and Wang [20], Guo and Zhang [21], Kawabe [24], Maghsoudi [33], Pap [39–42], Suzuki [45], Watanabe et al. [46], etc.).

The notion of non-atomicity for set functions also plays a key role in measure theory and its applications and extensions. For classical measures taking values in

A. Gavriluţ et al., *Atomicity through Fractal Measure Theory*,
https://doi.org/10.1007/978-3-030-29593-6_9

finite dimensional Banach spaces, it guarantees the connectedness of range. Even just replacing σ-additivity with finite additivity for measures requires some stronger non-atomicity property for the same conclusion to hold.

Because of their multiple applications, in game theory or mathematical economics, the study concerning atoms and non-atomicity for additive, respectively, non-additive set functions has developed. In this sense, we mention the contributions of Cavaliere and Ventriglia [5], Chiţescu [6, 7], Hart and Neyman [22], Khare and Singh [26], Klimkin and Svistula [27], Li and Mesiar [30], Li et al. [31, 32], Olejček [37, 38], Pap [39–42], Rao and Rao [44], Suzuki [45], Wu and Bo [48] and many others. In [11–15], non-atomicity and other related properties, like Darboux or Saks property were studied in the set-valued case, for set multifunctions taking values in the family of all nonempty closed subsets of a real normed space, family endowed with the Hausdorff-Pompeiu hypertopology.

In this chapter, we consider, more generally, the situation of non-additive set multifunctions taking values in the family of all nonempty subsets of a topological space, family endowed with an arbitrary hypertopology. We obtain results concerning atoms and non-atomicity by means of regularity which are also defined in this general setting. Thus, we generalize previous results of Pap [39, 40], Gavriluţ [11], Gavriluţ and Agop [16].

9.1 Basic notions, terminology and results

Let (X, τ) be a topological space, T an abstract nonvoid set, $\mathcal{C}$ a ring of subsets of T and $\mu : \mathcal{C} \to \mathcal{P}_0(X)$ an arbitrary set multifunction.

Suppose $\hat{\tau}$ is an arbitrary hypertopology (like Hausdorff topology $\hat{\tau}_H$, Vietoris topology $\hat{\tau}_V$, Wijsman topology $\hat{\tau}_W$, proximal topology $\hat{\tau}_P$ or any other) on $\mathcal{P}_0(X)$ (or its appropriate subfamily) and $\mathcal{B}_0$ is a base of vicinities of $\mu(\emptyset)$ in $\hat{\tau}$ (see, for details, Chapters 1 and 2).

The following notions are the natural extension to the general set-valued case of the corresponding notions from non-additive Measure Theory [40, 41, 45].

Let $\mu : \mathcal{C} \to \mathcal{P}_0(X)$ be an arbitrary set multifunction.

Definition 9.1.1

(i) A set $M \in \mathcal{C}$ is called an atom of μ if $\mu(M) \supsetneq \mu(\emptyset)$ and for every $N \in \mathcal{C}$, with $N \subseteq M$, we have $\mu(N) = \mu(\emptyset)$ or $\mu(M \backslash N) = \mu(\emptyset)$;

(ii) μ is called non-atomic (or atomless) if it has no atoms.

Definition 9.1.2 μ is called:

(i) monotone (or fuzzy) if $\mu(M) \subseteq \mu(N)$, for every $M, N \in \mathcal{C}$ with $M \subseteq N$;

(ii) null-additive if $\mu(M \cup N) = \mu(M)$, for every $M, N \in \mathcal{C}$, with $\mu(N) = \mu(\emptyset)$;

(iii) diffused if for every $t \in T$, whenever $\{t\} \in \mathcal{C}$, we have $\mu(\{t\}) = \mu(\emptyset)$.

The following statements are immediate.

Remark 9.1.3

(I) If X is a real normed space with the origin 0 and if $\mu(\emptyset) = \{0\}$, then one obtains the notions from [11–15]. Particularly, if $X = \mathbb{R}$, we get the notion of atom [39–42] and if one thinks about this notion, he observes that any part of an atom is either negligible, or it covers the entire atom. A physical interpretation is that the negligible part corresponds to the corpuscle (when the collapse occurs), while the part which covers the entire set (precisely, the atom) corresponds to the wave. Thus, it results the absolute dominance of one over the other. In this way, an atom can be considered as the elementary brick which includes all the properties of the "matter" which originated from.

(II) Suppose μ is monotone. Then:

(i) μ is non-atomic iff for every $M \in \mathcal{C}$ with $\mu(M) \supsetneq \mu(\emptyset)$, there is $N \in \mathcal{C}$, with $N \subsetneq M$, so that $\mu(N) \supsetneq \mu(\emptyset)$ and $\mu(M \backslash N) \supsetneq \mu(\emptyset)$.

(ii) If M is an atom of μ and if $N \in \mathcal{C}$, with $N \subseteq M$ is so that $\mu(N) \supsetneq \mu(\emptyset)$, then N is an atom of μ and $\mu(M \backslash N) = \mu(\emptyset)$.

Thus, any part of an atom is an atom, too and this yields to the non-differentiable and particularly, the fractal character of the physical phenomena in nature. In this way, holography is an expression of the fractal self-similarity property.

Let us note that significant progresses concerning fractality via Hausdorff and Vietoris topology have been recently obtained by Banakh and Novosad [3], Kunze *et. al* [29], Wicks [47], etc.

In physical theories there exist both the smallest scale (the Planck scale) and the largest scale (the cosmological scale). From here on, it can be built either the entire "zoology" of the elementary particles or the whole Universe (double stars, star clusters, galaxies, double galaxies, clusters of galaxies, etc.).

(III) If μ is null-additive, then μ is null-null-additive, i.e., $\mu(M \cup N) = \mu(\emptyset)$, for every $M, N \in \mathcal{C}$, with $\mu(M) = \mu(N) = \mu(\emptyset)$.

In what follows, unless stated otherwise in this chapter, $\mu : \mathcal{C} \to \mathcal{P}_0(X)$ is a *monotone null-additive set multifunction.*

9.2 Non-atomicity via regularity

In the sequel, let T be a locally compact Hausdorff space, $\mathcal{C}$ an arbitrary ring of subsets of T and $\mathcal{B}$ the Borel δ-ring. In what follows, we shall link the properties of atom/non-atomicity and regularity, which is an important property of continuity, connecting measure theory and topology. In this way, it approximates general Borel sets by more tractable sets, such as compact and or open sets and it can be considered as an approximation property but also as a continuity one with respect to a proper topology on $\mathcal{P}(T)$ (Dinculeanu [9], Ch. III, p. 197):

Let $\mathcal{K}$ be the family of all compact subsets of T and $\mathcal{O}$ the family of all open subsets of T. For every $K \in \mathcal{K}$ and every $O \in \mathcal{O}$, with $K \subseteq O$, we denote $\mathcal{I}(K, O) = \{M \subseteq T; K \subseteq M \subseteq O\}$. Since $\mathcal{I}(K, O) \cap \mathcal{I}(K', O') = \mathcal{I}(K \cup K', O \cap O')$, for every $\mathcal{I}(K, O), \mathcal{I}(K', O')$, the family $\{\mathcal{I}(K, O)\}_{\substack{K \in \mathcal{K} \\ O \in \mathcal{O}}}$ is a base of a topology $\tilde{\tau}$ on $\mathcal{P}(T)$.

$\tilde{\tau}$ also denotes the topology induced on any subfamily $\mathcal{S} \subseteq \mathcal{P}(T)$ of subsets of T.

$\tilde{\tau}_l$ (respectively, $\tilde{\tau}_r$) denotes the topology induced on $\{\mathcal{I}(K)\}_{K \in \mathcal{K}} = \{\{M \subseteq T; K \subseteq M\}\}_{K \in \mathcal{K}}$ (respectively, $\{\mathcal{I}(O)\}_{O \in \mathcal{O}} = \{\{M \subseteq T; M \subseteq O\}\}_{O \in \mathcal{O}}$) (Dinculeanu [9], Ch. III, p. 197–198).

Definition 9.2.1 A class $\mathcal{F} \subseteq \mathcal{P}(T)$ is dense in $\mathcal{P}(T)$ with respect to the topology induced by $\tilde{\tau}$ if for every $K \in \mathcal{K}$ and every $O \in \mathcal{O}$, with $K \subseteq O$, there is $M \in \mathcal{F}$ so that $K \subseteq M \subseteq O$.

Remark 9.2.2 The following remarks are easy to verify since T is locally compact.

(1) $\mathcal{B}$ is dense in $\mathcal{P}(T)$ with respect to the topology induced by $\tilde{\tau}$.
(2) (i) For every $M \in \mathcal{C}$, there is $O \in \mathcal{O} \cap \mathcal{C}$ so that $M \subseteq O$.
 (ii) If $\mathcal{C}$ is $\mathcal{B}$, then for every $M \in \mathcal{C}$, there are $K \in \mathcal{K} \cap \mathcal{C}$ and $O \in \mathcal{O} \cap \mathcal{C}$ so that $K \subseteq M \subseteq O$.
 In this way, the following notion of regularity has consistency on $\mathcal{B}$.

Definition 9.2.3 μ is (inner) regular on $\mathcal{C}$ if for every $M \in \mathcal{C}$ and every $B_0 \in \mathcal{B}_0$, there is $K = K(M, B_0) \in \mathcal{K} \cap \mathcal{C}$ so that $K \subseteq M$ and $\mu(M \backslash K) \in \mathcal{B}_0$.

Every set $K \in \mathcal{K} \cap \mathcal{C}$ is regular.

Remark 9.2.4

(i) In recent years, due to the development of computational graphics (for instance, in the automatic recognition of figures problems), it was necessary to measure accurately the matching, i.e., to calculate the distance between two sets of points. This led to the need to operate with an acceptable distance, which has to satisfy the first condition in the definition of a distance: the distance is zero if and only if the overlap is perfect. An appropriate metric in these issues is the Hausdorff metric on which we will refer in the following and which, roughly speaking, measures the degree of overlap of two compact sets.

 People recognize familiar faces in a similar way by using interior facial features (facial regions) such as eyes, nose, mouth and so on. In [25, 28, 34] it is proved that monotone (or fuzzy) measures are of particular interest with this regard given their monotonicity property (which stands in a clear contrast with the more restrictive additivity property inherent to probability—like measures).

 If $(X, ||\cdot||)$ is a real normed space with the origin 0 and $\mathcal{F}(X)$ the family of nonempty closed subsets of X, let h be the Hausdorff-Pompeiu pseudometric on $\mathcal{F}(X)$.

 For every $A \in \mathcal{F}(X)$, we denote $|A| = h(A, \{0\})$.

By $\hat{\tau}_H$, we denote the Hausdorff-Pompeiu topology on $\mathcal{F}(X)$. $\hat{\tau}_H$ is the supremum $\hat{\tau}_H^+ \vee \hat{\tau}_H^-$ of the upper Hausdorff topology $\hat{\tau}_H^+$ and the lower Hausdorff topology $\hat{\tau}_H^-$ (Hu and Papageorgiou [23]) ($\hat{\tau}_H^+$ has as a base of vicinities the family $\{U_+^H(A,\varepsilon)\}_{\varepsilon>0}$, where $U_+^H(A,\varepsilon) = \{B \in \mathcal{F}(X); B \subset S_\varepsilon(A)\}$ and $\hat{\tau}_H^-$ has as a base of vicinities the family $\{U_-^H(A,\varepsilon)\}_{\varepsilon>0}$, where $U_-^H(A,\varepsilon) = \{B \in \mathcal{F}(X); A \subset S_\varepsilon(B)\}$, where, generally, if $C \in \mathcal{F}(X)$, then $S_\varepsilon(C) = \{x \in X; \exists c \in C, d(x,c) < \varepsilon\}$.

A base of vicinities of 0 in $\hat{\tau}_H^+$ is $\{\{A \in \mathcal{F}(X); A \subseteq S_\varepsilon(\{0\})\}\}_{\varepsilon>0}$, where $S_\varepsilon(\{0\}) = \{x \in X; ||x|| < \varepsilon\}$.

Suppose $\mu(\emptyset) = \{0\}$.

Since for every $A \in \mathcal{F}(X)$, $A \subseteq S_\varepsilon(\{0\})$ iff $e(A,\{0\}) = |A| < \varepsilon$, then Definition 9.2.3 generalizes the notion of R_l'-regularity [11] in $\hat{\tau}_H^+$ (μ is R_l'-regular in $M \in \mathcal{C}$ if for every $\varepsilon > 0$, there exists $K = K(M,\varepsilon) \in \mathcal{K} \cap \mathcal{C}$ so that $K \subseteq M$ and $|\mu(M \backslash K)| < \varepsilon$).

(ii) If X is a Hausdorff linear topological space with the origin 0, we consider on $\mathcal{P}_0(X)$ the Vietoris topology $\hat{\tau}_V$ which is the supremum $\hat{\tau}_V^+ \vee \hat{\tau}_V^-$ of the upper Vietoris topology $\hat{\tau}_V^+$ and the lower Vietoris topology $\hat{\tau}_V^-$. A base for $\hat{\tau}_V^+$ is $\{A \in \mathcal{P}_0(X); A \subseteq D\}_{D \in \tau}$.

In consequence, the above definition also generalizes the notion of $R_l'^+$-regularity [16] in $\hat{\tau}_V^+$.

We observe that the topology $\tilde{\tau}$ restricted to $\mathcal{P}_0(T)$ having as a subbase the family $\{\mathcal{I}(K,O)\}_{K \in \mathcal{K} O \in \mathcal{O}}$ is in fact the Vietoris topology.

According to the above remarks, in the sequel we shall consider the following version of regularity:

Definition 9.2.5 μ is *(inner) regular on* $\mathcal{C}$ if for every $M \in \mathcal{C}$ and every $V \in \tau$ with $\mu(\emptyset) \subseteq V$, there is $K = K(M,V) \in \mathcal{K} \cap \mathcal{C}$ so that $K \subseteq M$ and $\mu(M \backslash K) \subseteq V$.

We now state our main results concerning atoms/non-atomicity via regularity. The following theorem generalizes Theorem 4.3.1 from Chapter 4.

Theorem 9.2.6 *If X is a T_1 space, $\mu : \mathcal{B} \to \mathcal{P}_0(X)$ is regular, $\mu(\emptyset)$ is a singleton and $M \in \mathcal{B}$ is an arbitrary set which satisfies the condition $\mu(M) \supsetneq \mu(\emptyset)$, then the following statements hold:*

(*i*) *M is an atom of μ iff there is a unique $m \in M$ so that $\mu(M \backslash \{m\}) = \mu(\emptyset)$.*
(*ii*) *If for every $t \in T$ there is $M_t \in \mathcal{B}$ so that $t \in M_t$ and $\mu(M_t) = \mu(\emptyset)$, then μ is non-atomic.*
(*iii*) *μ is non-atomic iff μ is diffused.*

Proof

(i) The *Only if part*.

Let $M \in \mathcal{B}$ be an arbitrary atom of μ and we consider the family $\mathcal{K}_M = \{K \subseteq M; K$ is a compact set and $\mu(M \backslash K) = \mu(\emptyset)\} \subseteq \mathcal{B}$.

$\mathcal{K}_M$ is nonvoid since if we suppose on the contrary that $\mathcal{K}_M = \emptyset$, then for every compact set $K \subseteq M$, we have $\mu(M \backslash K) \supsetneq \mu(\emptyset)$. Because M is an atom of μ, then $\mu(K) = \mu(\emptyset)$, so, according to the null-additivity of μ, $\mu(M) = \mu(M \setminus K)$. In consequence, $\mu(M) \supsetneq \mu(\emptyset)$. Now, because μ is regular, then for every $U \in \tau$, with $\mu(\emptyset) \subseteq U$ there is $K = K(M, U) \in \mathcal{K} \cap \mathcal{C}$ so that $K \subseteq M$ and $\mu(M \setminus K) \subseteq U$, whence $\mu(M) \subseteq U$, for every $U \in \tau$, with $\mu(\emptyset) \subseteq U$. Since X is $T_1$and $\mu(\emptyset)$ is a singleton, we finally get that $\mu(M) = \mu(\emptyset)$, which is a contradiction.

Now, we prove that every $K \in \mathcal{K}_M$ is an atom of μ. Indeed, if $K \in \mathcal{K}_M$, then $\mu(\emptyset) \subsetneq \mu(M) = \mu(K)$. Also, for every $N \in \mathcal{B}$, with $N \subseteq K$, since $K \subseteq M$ and M is an atom of μ, we get that $\mu(N) = \mu(\emptyset)$ or $\mu(M \setminus N) = \mu(\emptyset)$. If $\mu(M \setminus N) = \mu(\emptyset)$, then $\mu(\emptyset) \subseteq \mu(K \setminus N) \subseteq \mu(M \setminus N) = \mu(\emptyset)$, so $\mu(K \setminus N) = \mu(\emptyset)$.

In consequence, for every $N \in \mathcal{B}$, with $N \subseteq K$, we have $\mu(N) = \mu(\emptyset)$ or $\mu(K \setminus N) = \mu(\emptyset)$, whence K is indeed an atom of μ.

We note that for every $K_1, K_2 \in \mathcal{K}_M$, we have $K_1 \cap K_2 \in \mathcal{K}_M$. Indeed, if $K_1, K_2 \in \mathcal{K}_M$, then $K_1 \cap K_2$ is a compact set of T and $\mu(\emptyset) \subseteq \mu(M \setminus (K_1 \cap K_2)) = \mu((M \setminus K_1) \cup (M \setminus K_2)) = \mu(\emptyset)$, whence $\mu(M \setminus (K_1 \cap K_2)) = \mu(\emptyset)$.

We observe that $\bigcap_{K \in \mathcal{K}_M} K$ is nonvoid. If, on the contrary, $K_0 = \bigcap_{K \in \mathcal{K}_M} K = \emptyset$, then there are $K_1, K_2, \ldots, K_{n_0} \in \mathcal{K}_M$ so that $\bigcap_{i=1}^{n_0} K_i = \emptyset$, whence $\mu(\bigcap_{i=1}^{n_0} K_i) = \mu(\emptyset)$. Because $\bigcap_{i=1}^{n_0} K_i \in \mathcal{K}_M$, this implies $\mu(\bigcap_{i=1}^{n_0} K_i) \supsetneq \mu(\emptyset)$, a contradiction.

We prove that $K_0 \in \mathcal{K}_M$. It is evident that K_0 is compact and if $K \in \mathcal{K}_M$, then $\mu(M \setminus K) = \mu(\emptyset)$. We also observe the following:

If $K = K_0$, then $K_0 \in \mathcal{K}_M$ and if $K \neq K_0$, then $K_0 \subsetneq K$.

Because $\mu(\emptyset) \subseteq \mu(M \setminus K_0) = \mu((M \setminus K) \cup (K \setminus K_0)) = \mu(K \setminus K_0)$, it is sufficient to establish that $\mu(K \setminus K_0) = \mu(\emptyset)$. If on the contrary, $\mu(K \setminus K_0) \supsetneq \mu(\emptyset)$, let $N \in \mathcal{B}$ be such that $N \subseteq K \setminus K_0$. Then $N \subseteq K$ and, since K is an atom of μ, we have $\mu(N) = \mu(\emptyset)$ or $\mu(K \setminus N) = \mu(\emptyset)$.

If $\mu(K \setminus N) = \mu(\emptyset)$, then $\mu((K \setminus K_0) \setminus N) = \mu(\emptyset)$, hence $K \setminus K_0$ is an atom of μ.

Because M is an atom of μ and $\mu(K \setminus K_0) \supsetneq \mu(\emptyset)$, then $\mu(M \setminus (K \setminus K_0)) = \mu(\emptyset)$. Consequently, $\mathcal{K}_M = \{N \subseteq M; N \text{ is a compact set and } \mu(M \setminus N) = \mu(\emptyset)\}$ and $\mathcal{K}_{K \setminus K_0} = \{C \subseteq K \setminus K_0; C \text{ is compact and } \mu((K \setminus K_0) \setminus C) = \mu(\emptyset)\}$.

Let be $C \in \mathcal{K}_{K \setminus K_0}$. Then $\mu((K \setminus K_0) \setminus C) = \mu(\emptyset)$ and, since $\mu(M \setminus (K \setminus K_0)) = \mu(\emptyset)$, we get $\mu(\emptyset) \subseteq \mu(M \setminus C) = \mu(M \setminus (K \setminus K_0) \cup (K \setminus K_0) \setminus C) = \mu(\emptyset)$, hence $\mu(M \setminus C) = \mu(\emptyset)$, which implies that $C \in \mathcal{K}_M$. Therefore, $K_0 \subseteq C$. Since $C \subseteq K \setminus K_0$, we have the contradiction and, consequently, $\mu(K \setminus K_0) = \mu(\emptyset)$.

We now prove that K_0 is a singleton $\{m\}$. If on the contrary, there are $m, n \in M$, with $m \neq n$ and $K_0 \supseteq \{m, n\}$, since T is a locally compact Hausdorff space, there is an open neighborhood V of m so that $n \notin \overline{V}$.

Noting that $K_0 = (K_0 \backslash V) \cup (K_0 \cap \overline{V})$ and $K_0 \backslash V$, $K_0 \cap \overline{V}$ are nonvoid compact subsets of M, we prove that $K_0 \backslash V \in \mathcal{K}_M$ or $K_0 \cap \overline{V} \in \mathcal{K}_M$. If on the contrary $K_0 \backslash V \notin \mathcal{K}_M$ and $K_0 \cap \overline{V} \notin \mathcal{K}_M$, then $\mu(M \backslash (K_0 \backslash V)) \supsetneq \mu(\emptyset)$ and $\mu(M \backslash (K_0 \cap \overline{V})) \supsetneq \mu(\emptyset)$.

Since M is an atom of μ, $M \backslash (K_0 \backslash V) \subseteq M$ and $M \backslash (K_0 \cap \overline{V}) \subseteq M$, then $\mu(K_0 \backslash V) = \mu(\emptyset)$ and $\mu(K_0 \cap \overline{V}) = \mu(\emptyset)$. In consequence, $\mu(K_0) = \mu(\emptyset)$, which implies that $\mu(\emptyset) \subsetneq \mu(M) = \mu((M \backslash K_0) \cup K_0)) = \mu(\emptyset)$, which is a contradiction.

Therefore, $K_0 \backslash V \in \mathcal{K}_M$ or $K_0 \cap \overline{V} \in \mathcal{K}_M$. Because $K_0 \subseteq K$, for every $K \in \mathcal{K}_M$, we have $K_0 \subseteq K_0 \backslash V$ or $K_0 \subseteq K_0 \cap \overline{V}$, which is impossible.

The uniqueness is immediate. So, $\exists! m \in M$ such that $\mu(M \backslash \{m\}) = \mu(\emptyset)$ and thus $\mu(M) = \mu(\{m\})$.

The *If part.*

We consider $M \in \mathcal{B}$ with $\mu(M) \supsetneq \mu(\emptyset)$ and let be $N \in \mathcal{B}$ with $N \subseteq M$.

If $m \notin N$, then $N \subseteq M \backslash \{m\}$ and because $\mu(M \backslash \{m\}) = \mu(\emptyset)$, then $\mu(N) = \mu(\emptyset)$.

If $m \in N$, then $M \backslash N \subseteq M \backslash \{m\}$, hence $\mu(M \backslash N) = \mu(\emptyset)$. Consequently, M is an atom of μ.

(ii) If on the contrary, there is an atom $M_0 \in \mathcal{B}$ of μ, by i) $\exists! m \in M_0$ so that $\mu(M_0 \backslash \{m\}) = \mu(\emptyset)$. By the hypothesis, for $m \in M_0$ there is $\widetilde{M} \in \mathcal{B}$ such that $m \in \widetilde{M}$ and $\mu(\widetilde{M}) = \mu(\emptyset)$. Because $M_0 \subseteq (M_0 \backslash \{m\}) \cup \widetilde{M}$, we get $\mu(\emptyset) \subsetneq \mu(M_0) \subseteq \mu((M_0 \backslash \{a\}) \cup \widetilde{M}) = \mu(\emptyset)$, which is a contradiction and in consequence μ is non-atomic.

(iii) The *If part.*

The statement is immediate according to ii).

The *Only if part.*

If on the contrary there is $t_0 \in T$ so that $\mu(\{t_0\}) \supsetneq \mu(\emptyset)$, because μ is non-atomic there is a set $N \in \mathcal{B}$ such that $N \subseteq \{t_0\}$, $\mu(N) \supsetneq \mu(\emptyset)$ and $\mu(\{t_0\} \backslash N) \supsetneq \mu(\emptyset)$. In consequence, $N = \emptyset$ or $N = \{t_0\}$, which is false.

Remark 9.2.7

(I) The above theorem is a generalization of the results from [39, 40] obtained for non-additive real-valued set functions and also from [11] and [16] obtained for non-additive set multifunctions in Hausdorff topology, respectively, in Vietoris topology. We shall explain this fact in what follows:

In fact, Theorem 9.2.6 does not contain Theorem 1 from [39], i.e., the latter does not follow from the former. Indeed, every monotone single-valued set multifunction μ takes a constant value, i.e., $\mu(M) = \mu(\emptyset)$, for every $M \in \mathcal{C}$. Therefore, in this sense, Theorem 9.2.6 is not a generalization of Theorem 1 [39]. However, Theorem 9.2.6 can be considered to be a generalization of Theorem 1 [39] since the partial order on $\mathcal{P}_0(X)$ defined by the usual set order is a generalization of the order "$<$" on $[0, \infty)$ and one can use the "induced set multifunction" $\mu : \mathcal{C} \to \mathcal{P}_0([0, \infty))$, $\mu(M) = [0, m(M)]$, where $m : \mathcal{C} \to [0, \infty)$, with $m(\emptyset) = 0$.

(II) A measure is non-atomic iff the measure of any singleton from the space is zero. It means that in the construction of a physical theory, besides the "material" part, it should be also considered the vacuum condition of this matter as its complement.

References

1. Agahi, H., Mesiar, R., Ouyang, Y.: Further development of Chebyshev type inequalities for Sugeno integrals and T-(S-)evaluators. Kybernetika **46**(1), 83–95 (2010)
2. Apreutesei, G.: Families of subsets and the coincidence of hypertopologies. Ann. Alexandru Ioan Cuza Univ. Math. **XLIX**, 1–18 (2003)
3. Banakh, T., Novosad, N.: Micro and macro fractals generated by multi-valued dynamical systems. Fractals **22**(4), 1450012 (2014)
4. Beer, G.: Topologies on Closed and Closed Convex Sets. Kluwer Academic Publishers, Dordrecht (1993)
5. Cavaliere, P., Ventriglia, F.: On nonatomicity for non-additive functions. J. Math. Anal. Appl. **415**(1), 358–372 (2014)
6. Chiţescu, I.: Finitely purely atomic measures and $\mathcal{L}^p$-spaces. An. Univ. Bucureşti Şt. Natur. **24**, 23–29 (1975)
7. Chiţescu, I.: Finitely purely atomic measures: coincidence and rigidity properties. Rend. Circ. Mat. Palermo (2) **50**(3), 455–476 (2001)
8. Croitoru, A.: Convergences and topology via sequences of multifunctions. Infor. Sci. **282**, 250–260 (2014)
9. Dinculeanu, N.: Measure Theory and Real Functions (in Romanian). Ed. Did. şi Ped., Bucureşti (1964)
10. Gagolewski, M., Mesiar, R.: Monotone measures and universal integrals in a uniform framework for the scientific impact assessment problem. Infor. Sci. (263), 166–174 (2014)
11. Gavriluţ, A.: Non-atomicity and the Darboux property for fuzzy and non-fuzzy Borel/Baire multivalued set functions. Fuzzy Sets Syst. **160**, 1308–1317 (2009). Erratum in Fuzzy Sets Syst. **161**, 2612–2613 (2010)
12. Gavriluţ, A., Croitoru, A.: Non-atomicity for fuzzy and non-fuzzy multivalued set functions. Fuzzy Sets Syst. **160**(14), 2106–2116 (2009)
13. Gavriluţ, A.: Fuzzy Gould integrability on atoms. Iran. J. Fuzzy Syst. **8**(3), 113–124 (2011)
14. Gavriluţ, A., Croitoru, A.: On the Darboux property in the multivalued case. Ann. Univ. Craiova Math. Comp. Sci. Ser. **35**, 130–138 (2008)
15. Gavriluţ, A., Croitoru, A.: Pseudo-atoms and Darboux property for set multifunctions. Fuzzy Sets Syst. **161**(22), 2897–2908 (2010)
16. Gavriluţ, A., Agop, M.: Approximation theorems for fuzzy set multifunctions in Vietoris topology. Physical implications of regularity. Iran. J. Fuzzy Syst. **12**(1), 27–42 (2015)
17. Gavriluţ, A.C.: A Lusin type theorem for regular monotone uniformly autocontinuous set multifunctions. Fuzzy Sets Syst. **161**(22), 2909–2918 (2010)
18. Gavriluţ, A.C.: On the regularities of fuzzy set multifunctions with applications in variation, extensions and fuzzy set-valued integrability problems. Inform. Sci. **224**, 130–142 (2013)
19. Gavriluţ, A.C.: Remarks on monotone interval-valued set multifunctions. Infor. Sci. **259**, 225–230 (2014)
20. Gong, Z., Wang, L.: The Henstock Stieltjes integral for fuzzy-number-valued functions. Inform. Sci. **188**, 276–297 (2012)
21. Guo, C., Zhang, D.: On set-valued fuzzy measures. Inform. Sci. **160**(14), 13–25 (2004)
22. Hart, S., Neyman, A.: Values of non-atomic vector measure games: are they linear combinations of the measures? J. Math. Econ. **17**(1), 31–40 (1988)

23. Hu, S., Papageorgiou, N.S.: Handbook of Multivalued Analysis, Vol. I. Kluwer Academic Publishers, Dordrecht (1997)
24. Kawabe, J.: Regularity and Lusin's theorem for Riesz space-valued fuzzy measures. Fuzzy Sets Syst. **158**(8), 895–903 (2007)
25. Karczmarek, P., Pedrycz, W., Reformat, M., Akhoundi, E.: A study in facial regions saliency: a fuzzy measure approach. Soft Comput. **18**(2), 379–391 (2014)
26. Khare, M., Singh, A.K.: Atoms and Dobrakov submeasures in effect algebras. Fuzzy Sets Syst. **159**(9), 1123–1128 (2008)
27. Klimkin, V.M., Svistula, M.G.: Darboux property of a non-additive set function. Mat. Sb. **192**(7), 41–50 (2001)
28. Ko, H., Bae, K., Choi, J., Kim, S.H., Choi, J.: Similarity recognition using context-based pattern for cyber-society. Soft Comput. **20**(11), 4565–4573 (2016). https://doi.org/10.1007/s00500-015-1763-9
29. Kunze, H., la Torre, D., Mendivil, F., Vrscay, E.R.: Fractal-Based Methods in Analysis. Springer, Berlin (2012)
30. Li, J., Mesiar, R.: Lusin's theorem on monotone measure spaces. Fuzzy Sets Syst. **175**, 75–86 (2011)
31. Li, J., Mesiar, R., Pap, E.: Atoms of weakly null-additive monotone measures and integrals. Inform. Sci. **257**, 183–192 (2014)
32. Li, J., Yasuda, M., Song, J.: Regularity properties of null-additive fuzzy measure on metric spaces. Lecture Notes in Artificial Intelligence, vol. 3558, pp. 59–66. Springer, Berlin (2005)
33. Maghsoudi, S.: Certain strict topologies on the space of regular Borel measures on locally compact groups. Topol. Appl. **160**(14), 1876–1888 (2013)
34. Marques, I., Graña, M.: Face recognition with lattice independent component analysis and extreme learning machines. Soft Comput. **16**(9), 1525–1537 (2012)
35. Mesiar, R., Li, J., Pap, E.: Superdecomposition integrals. Fuzzy Sets Syst. **259**, 3–11 (2015). https://doi.org/10.1016/j.fss.2014.05.003
36. Mesiar, R., Li, J., Pap, E.: The Choquet integral as Lebesgue integral and related inequalities. Kybernetika **46**(6), 1098–1107 (2010)
37. Olejček, V.: Darboux property of regular measures. Mat. Cas. **24**(3), 283–288 (1974)
38. Olejček, V.: Fractal construction of an atomic Archimedean effect algebra with non-atomic subalgebra of sharp elements. Kybernetika **48**(2), 294–298 (2012)
39. Pap, E.: Regular Borel t-decomposable measures. Univ. u Novom Sadu, Zb. Rad. Prirod. - Mat. Fak., Ser. Mat. **20**(2), 113–120 (1990)
40. Pap, E.: The range of null-additive fuzzy and non-fuzzy measures. Fuzzy Sets Syst. **65**(1), 105–115 (1994)
41. Pap, E.: Null-additive Set Functions. Mathematics and Its Applications, vol. 337. Springer, Berlin (1995)
42. Pap, E.: Some elements of the classical measure theory. In: Handbook of Measure Theory, pp. 27–82. Springer, Berlin (2002)
43. Pap, E., Strboja, M., Rudas, I.: Pseudo-Lp space and convergence. Fuzzy Sets Syst. **238**, 113–128 (2014)
44. Pap, E., Gavriluţ, A., Agop, M.: Atomicity via regularity for non-additive set multifunctions. Soft Comput. **20**(12), 4761–4766 (2016). https://doi.org/10.1007/s00500-015-20121-x
45. Rao, K.P.S.B., Rao, M.B.: Theory of Charges. Academic Press, New York (1983)
46. Suzuki, H.: Atoms of fuzzy measures and fuzzy integrals. Fuzzy Sets Syst. **41**, 329–342 (1991)
47. Watanabe, T., Kawasaki, T., Tanaka, T.: On a sufficient condition of Lusin's theorem for non-additive measures that take values in an ordered topological vector space. Fuzzy Sets Syst. **194**, 66–75 (2012)
48. Wu, C., Bo, S.: Pseudo-atoms of fuzzy and non-fuzzy measures. Fuzzy Sets Syst. **158**, 1258–1272 (2007)

Chapter 10
Extended atomicity through non-differentiability and its physical implications

10.1 Introduction

In this chapter, the mathematical concept of atomicity (and, particularly, that of minimal atomicity) is extended, based on the non-differentiability of the motion curves associated to the motions of the structural units of a complex system on a fractal manifold. For this purpose, firstly, different results concerning standard atomicity (and, particularly, minimal atomicity) are obtained from the mathematical procedure of the Quantum Measure Theory. Also, several physical implications are analyzed in the form of maximally entangled states coherence and decoherence through Young type experiments, etc. Further, based on the remark that Quantum Mechanics is a particular case of Fractal Mechanics at a specified scale resolution, we introduce the concept of fractal atomicity (and, particularly, that of fractal minimal atomicity) and we give some of their mathematical properties. In such an approach, one operates with an invariance group, with 1-forms and 2-forms which are absolutely invariant with respect to the invariance group, with the elementary measure of the invariance group, with parallel transport in the Levi-Civita sense on a hyperbolic manifold, etc. All these imply "synchronization" among the structural units of a complex system, circumstances left unspecified in an experiment in Jaynes sense, compactification of the angular moment in the null vectors space in the form of the spin, etc.

Measure Theory concerns with assigning a notion of size to sets. In the last years, non-additive measures theory was given an increasing interest due to its various applications in a wide range of areas (economics, social sciences, biology, philosophy, etc.). It is used to describe situations concerning conflicts or cooperations among intelligent rational players, giving an appropriate mathematical framework to predict the outcome of the process. Precisely, theories dealing with (pseudo)atoms and monotonicity are used in statistics, game theory, probabilities, artificial intelligence. The notion of non-atomicity for set (multi)functions plays a key role in Measure Theory and its applications and extensions. For classical

A. Gavriluţ et al., *Atomicity through Fractal Measure Theory*,
https://doi.org/10.1007/978-3-030-29593-6_10

measures taking values in finite dimensional Banach spaces, it guarantees the connectedness of range. Even just replacing σ-additivity with finite additivity for measures requires some stronger non-atomicity property for the same conclusion to hold.

Because of their multiple applications, in game theory or mathematical economics, the study concerning atoms and non-atomicity for additive, respectively, non-additive set functions has developed. Particularly, (non)atomic measures and purely atomic measures have been investigated (in different variants) due to their special form and their special properties (Chiţescu [10, 11], Cavaliere and Ventriglia [9], Gavriluţ and Agop [18], Gavriluţ and Croitoru [19–21], Gavriluţ [15–17], Gavriluţ, Iosif and Croitoru [22], Khare and Singh [34], Li et al. [35, 36], Pap [46–48], Pap et al. [49], Rao and Rao [51], Suzuki [60], Wu and Bo [61]).

Thus, one important application of Measure Theory is in probability, where a measurable set is interpreted as an event and its measure as the probability that the event will occur. Since probability is an important notion in Quantum Mechanics, Measure Theory's techniques could be used to study quantum phenomena. Unfortunately, one of the foundational axioms of Measure Theory does not remain valid in its intuitive application to Quantum Mechanics, since additivity fails in this framework.

Also, interesting results can be found in [1, 8, 23]. Although classical measure theory imposes strict additivity conditions, a rich theory of non-additive measures developed. Precisely, modifications of traditional Measure Theory (Pap [46–48]) led to Quantum Measure Theory (Gudder [25–29], Salgado [52], Sorkin [54–56], Surya and Waldlden [57]). Practically, an extended notion of a measure has been introduced and its applications to the study of interference, probability, and spacetime histories in Quantum Mechanics have been discussed (Schweizer and Sklar [53]).

Quantum Measure Theory is a generalization of Quantum Theory where physical predictions are computed from a matrix known as decoherence functional. Introduced by Sorkin in [54–56], quantum measures help us to describe Quantum Mechanics and its applications to Quantum Gravity and Cosmology (Hartle [30, 31], Phillips [50]). Quantum Measure Theory indicates a wide variety of applications, its mathematical structure being used in the standard quantum formalism.

Despite the continuous efforts of numerous scientists, reconciling General Relativity with Quantum Theory remains one of the most important open problems in Physics. The framework of general relativity suggests that one promising approach to such a unification will be by means of a reformulation of Quantum Theory in terms of histories rather than states. Following this idea, Sorkin [54–56], has proposed a history-based framework, which can accommodate both standard Quantum Mechanics as well as physical theories beyond the quantum formalism.

As we shall prove in this chapter, in such framework, Schrödinger's equation from Quantum Mechanics can be identified with a particular type of geodesic of the fractal space. In consequence, fundamental concepts of Quantum Mechanics can be extended to similar concepts, but on fractal manifolds. The aim of this chapter is to provide the mathematical-physical framework that is necessary to extend some of these concepts. Precisely, we extend the concept of atoms/pseudo-atoms to the concept of fractal minimal atom/fractal pseudo-atom, respectively. We also give characterizations from a mathematical viewpoint to these new concepts and we make explicit certain physical implications. The notion of a fractal minimal atom as a particular case of fractal atom is also discussed.

Firstly, we introduce the notion of an atom/pseudo-atom/minimal atom from a mathematical perspective and we give some of their mathematical properties. In this framework, we are looking for a physical correspondence in the Quantum Mechanics context. Secondly—taking into account that Quantum Mechanics is a particular case of Fractal Mechanics (for Peano curves at Compton scale resolution—see Section 10.4 of this chapter), situation when the Schrödinger equation identifies with a particular geodesic of a fractal manifold—we apply an inverse method. We make explicit the fundamental elements of Fractal Mechanics and starting from here we introduce new concepts as fractal atom, fractal pseudo-atom, etc.

10.2 Towards Quantum Measure Theory by means of Fractal Mechanics

The basic idea behind Quantum Measure Theory, or Generalized Quantum Mechanics, for that matter, is to provide a description of the world in terms of histories. A history is a classical description of the system under consideration for a given period of time, finite or infinite. If we are trying to describe a system of N particles, then a history will be given by N classical trajectories. If we are working with a field theory, then a history will correspond to the spatial configuration of the field as a function of time. In either case, Quantum Measure Theory tries to provide a way to describe the world through classical histories by extending the notion of probability theory which is clearly not rich enough to model our universe.

On the other hand, structures, self-structures, etc. of the nature can be assimilated to complex systems, taking into account both their functionality, as well as their structure (Mitchell [43], Nottale [44]). The models commonly used to study the dynamics of complex systems are based on the assumption, otherwise unjustified, of the differentiability of the physical quantities that describe it, such as density, momentum, energy, etc. (for mathematical models and for applications, see Mercheş and Agop [39], Nottale [44]).

The success of differentiable models must be understood sequentially, i.e. on domains large enough that differentiability and integrability are valid.

But differential method fails when facing the physical reality, with non-differentiable or non-integral physical dynamics, such as instabilities in the case of dynamics of complex systems, instabilities that can generate both chaos and patterns.

In order to describe such dynamics of complex systems, but still remaining tributary to a differential hypothesis, it is necessary to introduce, in an explicit manner, the scale resolution in the expressions of the physical variables that describe these dynamics and, implicitly, in the fundamental equations of “evolution” (for example, density, momentum, energy equations, etc.). This means that any dynamic variable, dependent, in a classical meaning, on both spatial coordinates and time (Michel and Thomas [41], Mitchell [43]), becomes, in this new context, dependent also on the resolution scale.

In other words, instead of working with a dynamic variable, described through a strictly non-differentiable mathematical function, we will just work with different approximations of that function, derived through its averaging at different resolution scales. Consequently, any dynamic variable acts as the limit of a functions family, the function being non-differentiable for a null resolution scale and differentiable for a non-zero resolution scale.

This approach, well adapted for applications in the field of dynamics of complex systems, where any real determination is conducted at a finite resolution scale, clearly implies the development both of a new geometric structure and of a physical theory (applied to dynamics of complex systems) for which the motion laws, invariant to spatial and temporal coordinates transformations, are integrated with scale laws, invariant at scale transformations.

Such a theory that includes the geometric structure based on the above presented assumptions was developed in the Scale Relativity Theory (Nottale [44]) and more recently in the Scale Relativity Theory with an arbitrary constant fractal dimension (Mercheş and Agop [39]). Both theories define the “fractal physics models” class (Mercheş and Agop [39], Nottale [44]).

Various theoretical aspects and applications of the Scale Relativity Theory with an arbitrary constant fractal dimension in the field of physics are presented in Mercheş and Agop [39], Nottale [44]. In this model, if we assume that the complexity of interactions in the dynamics of complex systems is replaced by non-differentiability, then the motions constrained on continuous, but differentiable curves in an Euclidean space are replaced with free motions, without any constrains, on continuous, but non-differentiable curves (fractal curves) in a fractal space. In other words, for time resolution scale that prove to be large when compared with the inverse of the highest Lyapunov exponent (Mandelbrot [37]), the deterministic trajectories are replaced by a collection of potential routes, so that the concept of “definite positions” is substituted by that of an ensemble of positions having a definite probability density (Mandelbrot [37], Mercheş and Agop [39], Nottale [44]).

In consequence, the motion curves have double identity: both geodesics of the fractal space and streamlines of a fractal fluid, whose entities (the structural units of the complex system) are substituted with the geodesics themselves so that any external constrains are interpreted as a selection of geodesics by means of measuring device.

Since in such conjecture the Quantum Mechanics becomes a particular case of Fractal Mechanics (for structural units movements of a complex system on Peano curves at Compton scale resolution—see Section 10.4 from this chapter), then Quantum Measure Theory could also become, in our opinion, a particular type of a Fractal Measure Theory (see Section 10.5).

10.3 Types of atoms in the mathematical approach

10.3.1 Atoms and pseudo-atoms

Let T be an abstract nonvoid set, $\mathcal{C}$ a ring of subsets of T and $(V, +, \cdot)$ a real linear space with the origin 0.

Definition 10.3.1 (Gavriluţ and Agop [18]) Let $m : \mathcal{C} \to V$ be a set function, with $m(\emptyset) = 0$.

(I) m is said to be:

(i) finitely additive (or, grade-1-additive) if $m(\bigcup_{i=1}^{n} A_i) = \sum_{i=1}^{n} m(A_i)$, for any arbitrary pairwise disjoint sets $(A_i)_{i\in\{1,2,\ldots,n\}} \subset \mathcal{C}$, $n \in \mathbb{N}^*$;

(ii) a grade-2-measure if

$$m(A{\cup}B{\cup}C){+}m(A){+}m(B){+}m(C) = m(A{\cup}B){+}m(B{\cup}C){+}m(A{\cup}C)(**)$$

for any pairwise disjoint sets $A, B, C \in \mathcal{C}$;

(II) Two sets $A, B \in \mathcal{C}$ are called m-compatible (denoted by AmB) if

$$m(A \cup B) + m(A \cap B) = m(A) + m(B)(*).$$

(i.e., m-compatible sets are those two sets for which the set function m behaves like a grade-1-measure);

(III) An arbitrary fixed set $A \in \mathcal{C}$ that is m-compatible with any set $B \in \mathcal{C}$ is said to be a macroscopic set.

Remark 10.3.2 (Gavriluţ and Agop [18])

(i) Some quantum objects interfere with each other, but others do not. Consequently, one can justify the name of a "macroscopic set" by the fact that it does not interfere with any set and thus it behaves like a non-quantum object in the macroscopic world.

(ii) One can immediately verify that the relation given by m-compatibility is reflexive, symmetric but it is not transitive.

(iii) Evidently, if m is grade-1-additive, then it is also a grade-2-measure, but the converse is not valid.

(iv) If $A \in \mathcal{C}$ is arbitrary, then A and $\emptyset$ are m-compatible.

(v) Suppose t_i, where $i \in \{1, 2, \ldots, n\}, n \in \mathbb{N}^*$ represent quantum objects or quantum events and let be their collection $T = \{t_1, t_2, \ldots, t_n\}$. One can need an interpretation of a "measure" on T, in situations when the additivity condition $(**)$ is not fulfilled:

Example 10.3.3 (Gavriluţ and Agop [18])

(i) Suppose ($T = \{t_1, t_2, \ldots, t_n\}, m : \mathcal{P}(T) \to \mathbb{R}_+$) is a finite measure space, where for every $i \in \{1, 2, \ldots, n\}$, t_i represents the particle, and m the mass. The real-valued set function associated to the mass is additive in the macroscopic world. On the quantum scale, these statements do not remain valid due to the annihilation and binding energy effects. For instance, if t_1 and t_2 represent an electron and a positron, respectively, then $m(\{t_1\}) = m(\{t_2\}) = 9,11\times 10^{-31}$ kg, but $m(\{t_1, t_2\}) = m(\{t_1\} \cup \{t_2\}) = 0$.

(ii) In Quantum Mechanics, the wave-particle duality principle states that every fermion (matter particle) and boson (force-carrying particle) is described by a wavefunction, i.e., a time-varying function, giving the particle's probability density at each point in space. These wavefunctions often behave like classical waves, exhibiting properties such as diffraction and interference.

The two-slit experiment proved that a beam of electrons shot through two narrow slits produces an interference pattern which is identical to the interference patterns produced by electromagnetic (light) waves. Thus, in situations involving particles, additivity of measures fails when interference occurs. Precisely, interference allows the union of sets of measure zero to have nonzero measure.

These are some reasons for in what follows we introduce several notions, weaker than classical additivity and also than those from Definition 10.3.1(i), (ii):

Definition 10.3.4 (Gavriluţ and Agop [18]) A set function $m : \mathcal{C} \to V$, with $m(\emptyset) = 0$, is said to be:

(i) null-additive^1 if $m(A \cup B) = m(A)$, for every disjoint $A, B \in \mathcal{C}$, with $m(B) = 0$;
(ii) null-additive^2 if $m(A \cup B) = m(A)$, for every $A, B \in \mathcal{C}$, with $m(B) = 0$;
(iii) null-null-additive if $m(A \cup B) = 0$, for every $A, B \in \mathcal{C}$, with $m(A) = m(B) = 0$;
(iv) null-equal if $m(A) = m(B)$, for every $A, B \in \mathcal{C}$, with $m(A \cup B) = 0$;
(v) a quantum measure (q-measure, for short) if it is a null-additive^1 and null-equal grade-2-measure;
(vi) diffused if $m(\{t\}) = 0$, whenever $\{t\} \in \mathcal{C}$.

Definition 10.3.5 (Gavriluţ et al. [22]) If V is, moreover, a Banach lattice, a set function $m : \mathcal{C} \to V$, with $m(\emptyset) = 0$, is said to be:

(i) null-monotone if for every $A, B \in \mathcal{C}$, with $A \subseteq B$, if $m(B) = 0$, then $m(A) = 0$;
(ii) monotone (or, fuzzy) if $m(A) \leq m(B)$, for every $A, B \in \mathcal{C}$, with $A \subseteq B$;
(iii) a submeasure (in the sense of Drewnowski [13]) if m is monotone and subadditive, i.e., $m(A \cup B) \leq m(A) + \nu(B)$, for every (disjoint) $A, B \in \mathcal{C}$;
(iv) σ-additive (or, a (vector) measure) if $m(\bigcup_{n=1}^{\infty} A_n) = \lim_{n\to\infty} \sum_{k=1}^{n} m(A_k)$, for every pairwise disjoint sets $(A_n)_{n\in\mathbb{N}^*} \subset \mathcal{C}$, with $\bigcup_{n=1}^{\infty} A_n \in \mathcal{C}$.

The notion of an algebra of sets is significant in measure theory since it contains conditions on which sets are measurable:

Definition 10.3.6 If $\mathcal{A}$ is an arbitrary σ-algebra of T and if $m : \mathcal{A} \to \mathbb{R}_+$ is a measure on $\mathcal{A}$, with $m(T) = 1$, then:

(i) The space $(T, \mathcal{A}, m)$ is said to be a sample space and m is said to be a probability measure;
(ii) The elements of T are called sample points or outcomes and the elements of $\mathcal{A}$ are called events.

In this case, for every $A \in \mathcal{A}$, $m(A)$ is interpreted as the probability of the event A to occur.

Remark 10.3.7

(i) The notion of a null-equal-measure has the following physical interpretation (Gavriluţ and Agop [18]): in the situation involving destructive interference, in order for two waves to produce complete destructive interference, thereby "cancelling out" each other, their original amplitudes must have been equal.
(ii) Any positive real-valued finitely additive set function $m : \mathcal{C} \to \mathbb{R}_+$ is a q-measure.
(iii) If $m(T) > 0$, then one can immediately generate a probability measure by means of a normalization process.

Remark 10.3.8

(I) (i) One observes that a set function $m : \mathcal{C} \to V$ is diffused if the measure of any singleton of the space is null. This means in the construction of a physical theory, the vacuum condition of the matter should be considered as its complement.
(ii) By [5], Shannon's entropy is a subadditive real-valued set function.
(iii) If V is a Banach lattice, $T = \{t_1, t_2, \ldots, t_n\}, n \in N^*$ is an arbitrary finite metric space and $m : \mathcal{P}(T) \to V$ (or, more general, if T is a T_1-separated topological space, $\mathcal{B}$ is the Borel σ-algebra of T generated by the lattice of all compact subsets of T and $m : \mathcal{B} \to V$) is null-additive and diffused, then $m(T) = 0$ (i.e., the space T is composed of particles which annihilate one each other).

(II) If $m : \mathcal{C} \to V$ is null-additive^1, then any two disjoint sets $A, B \in \mathcal{C}$, with $m(B) = 0$, are m-compatible.

(III) If $m : \mathcal{C} \to V$ is null-monotone, then:

(i) m is null-additive^1 if and only if it is null-additive^2. In this case, m will be simply called null-additive.
(ii) If m is null-null-additive, then it is null-equal.

Definition 10.3.9 (Gavriluţ and Croitoru [19–21]) Let $m : \mathcal{C} \to \mathbb{R}_+$ be a set function, with $m(\emptyset) = 0$.

(i) A set $A \in \mathcal{C}$ is said to be an atom of ν if $m(A) > 0$ and for every $B \in \mathcal{C}$, with $B \subseteq A$, we have $m(B) = 0$ or $m(A \backslash B) = 0$ (in a certain sense, an atom can be interpreted as being a black hole);
(ii) m is said to be non-atomic if it has no atoms (i.e., for every $A \in \mathcal{C}$ with $m(A) > 0$, there exists $B \in \mathcal{C}$, $B \subseteq A$, such that $m(B) > 0$ and $m(A \backslash B) > 0$);
(iii) A set $A \in \mathcal{C}$ is called a pseudo-atom of ν if $m(A) > 0$ and $B \in \mathcal{C}$, $B \subseteq A$ implies $m(B) = 0$ or $m(B) = m(A)$;
(iv) m is said to be non-pseudo-atomic if it has no pseudo-atoms (i.e., for every $A \in \mathcal{C}$ with $m(A) > 0$, there exists $B \in \mathcal{C}$, $B \subseteq A$, such that $m(B) > 0$ and $m(A) \neq m(B)$);
(v) m is said to be finitely purely atomic if there is a finite family $(A_i)_{i \in \{1,2,\ldots,n\}}$ of pairwise disjoint atoms of m so that $T = \bigcup_{i=1}^{n} A_i$ (in this case, the space T is a finite collection of pairwise disjoint atoms).

Remark 10.3.10

(i) One could think to the following physical interpretation of an atom/ a pseudo-atom: when the collapse occurs, the negligible part corresponds to the corpuscle, while the part which covers the entire set (precisely, the atom) corresponds to the wave. In this way, the absolute dominance of one over the other results, so an atom can be considered as the "elementary bridge" which includes all the properties of the "matter" (in the form of the corpuscle and of the wave) it originated form. More precisely, in our opinion, we discuss here about an Einstein-Podolsky-Rosen (EPR) type bridge (Susskind [58, 59]), in which the corpuscle and the wave are connected and can be interpreted as maximally entangled states of the same "physical objects".
(ii) In fact, atoms are singularities of the space-matter metric that will be in detail discussed in the last section.

We now recall the following properties concerning operations with atoms/pseudo-atoms:

Proposition 10.3.11 (Gavriluţ and Croitoru [19–21]) *Let be $m : \mathcal{C} \to \mathbb{R}_+$, with $m(\emptyset) = 0$.*

(i) (self-similarity of atoms) If m is null-monotone, $A \in \mathcal{C}$ is an atom of m and $B \in \mathcal{C}$, $B \subseteq A$ is such that $m(B) > 0$, then B is an atom of m and $m(A \backslash B) = 0$.
(ii) (self-similarity of pseudo-atoms) If $A \in \mathcal{C}$ is a pseudo-atom of and $B \in \mathcal{C}$, $B \subseteq A$ is such that $m(B) > 0$, then B is a pseudo-atom of m and $m(B) = m(A)$.
(iii) If $A, B \in \mathcal{C}$ are pseudo-atoms of and $m(A \cap B) > 0$, then $A \cap B$ is a pseudo-atom of m and $m(A \cap B) = m(A) = m(B)$.
(iv) Let $m : \mathcal{C} \to \mathbb{R}_+$ be null-additive and let $A, B \in \mathcal{C}$ be pseudo-atoms of m.

1. If $m(A \cap B) = 0$, then $A \backslash B$ and $B \backslash A$ are pseudo-atoms of m and $m(A \backslash B) = m(A)$, $m(B \backslash A) = m(B)$.

2. *If* $m(A) \neq m(B)$, *then* $m(A \cap B) = 0$, $m(A \backslash B) = m(A)$ *and* $m(B \backslash A) = m(B)$.

(v) *Let* $m : \mathcal{C} \to \mathbb{R}_+$ *be null-additive and let* $A, B \in \mathcal{C}$ *be pseudo-atoms of* m. *If* $m(A \cap B) > 0$ *and* $m(A \backslash B) = m(B \backslash A) = 0$, *then* $A \cap B$ *is a pseudo-atom of and* $m(A \Delta B) = 0$.

The following statements easily follow:

Proposition 10.3.12 *Suppose* $m : \mathcal{C} \to \mathbb{R}_+$ *is so that* $m(\emptyset) = 0$.

(i) *If* m *is finitely additive, then* $A \in \mathcal{C}$ *is an atom of* m *if and only if* A *is a pseudo-atom of* m.
(ii) *Any* $\{t\} \subseteq T$, *provided* $\{t\} \in \mathcal{C}$ *and* $m(\{t\}) > 0$, *is an atom of* m.
(iii) *If* m *is null-additive*[1], *then every atom of* m *is also a pseudo-atom. The converse is not generally valid.*

Example 10.3.13 Let $T = \{t_1, t_2\}$ be a finite abstract space composed of two elements.

(i) We consider the set function $m : \mathcal{P}(T) \to \mathbb{R}_+$ defined for every $A \subset T$ by

$$m(A) = \begin{cases} 2, & A = T \\ 1, & A = \{t_1\} \\ 0, & A = \{t_2\} \text{ or } A = \emptyset. \end{cases}$$

Then T is an atom and it is not a pseudo-atom of m.

(ii) We define $m : \mathcal{P}(T) \to \mathbb{R}_+$ by $m(A) = \begin{cases} 1, & A \neq \emptyset \\ 0, & A = \emptyset \end{cases}$, for every $A \subset T$.

Then m is null-additive and $T = \{t_1, t_2\}$ is a pseudo-atom of m, but it is not an atom.

Proposition 10.3.14 *If* $m : \mathcal{C} \to \mathbb{R}_+$ *is null-monotone and null-additive and if* $A \cup B$ *is an atom of* m, *then* A, B *are* m*-compatible (thus, any two components of an atom must be compatible).*

Proof

1. If $m(A) = 0$, then $m(A \cap B) = 0$ and $m(A \cup B) = m(B)$, so the conclusion follows.
2. If $m(A) > 0$, then by Proposition 10.3.11(i) A is an atom, too and $m((A \cup B) \backslash A) = m(B \backslash A) = 0$. Since $B = (B \backslash A) \cup (B \cap A)$, then $m(B) = m(A \cap B)$ and since $A \cup B = A \cup (B \backslash A)$, we get $m(A \cup B) = m(A)$.

Definition 10.3.15 If $m : \mathcal{C} \to \mathbb{R}_+$ is so that $m(\emptyset) = 0$, we consider the variation of m, $\overline{m} : \mathcal{P}(T) \to [0, \infty]$, defined for every $A \in \mathcal{P}(T)$ by:

$$\overline{m}(A) = \sup \left\{ \sum_{i=1}^{n} m(A_i);\ A = \bigcup_{i=1}^{n} A_i,\ A_i \in \mathcal{C},\ \forall i = \overline{1, n},\ A_i \cap A_j = \emptyset,\ i \neq j \right\}.$$

We say that m is of finite variation if $\overline{m}(T) < \infty$.

In what follows, we give some examples of q-measures. Other examples will be given in Section 10.3.3.

Proposition 10.3.16 *If* $m : \mathcal{C} \to \mathbb{R}_+$ *is a submeasure of finite variation, then* $\overline{m}$ *is a q-measure.*

Proof Since m is a submeasure of finite variation, then, according to Dinculeanu [12], $\overline{m} : \mathcal{C} \to [0, \infty)$ is finitely additive, so it is a q-measure.

In what follows, let $\mathcal{K}$ be the lattice of all compact subsets of a locally compact Hausdorff space T and $\mathcal{B}$ be the Borel σ-algebra generated by $\mathcal{K}$. In such framework, the following definition is consistent:

Definition 10.3.17 (Pap [46–48]) $m : \mathcal{B} \to \mathbb{R}_+$ is said to be regular if for every $A \in \mathcal{B}$ and every $\varepsilon > 0$, there exist $K \in \mathcal{K}$ and an open set $D \in \mathcal{B}$ such that $K \subset A \subset D$ and $m(D \backslash K) < \varepsilon$.

Theorem 10.3.18 (Pap [46–48]) *Suppose* $m : \mathcal{B} \to \mathbb{R}_+$ *is a monotone null-additive regular set function. If* $A \in \mathcal{B}$ *is an atom of* m*, there exists a unique point* $a \in A$ *so that* $m(A\backslash\{a\}) = 0$ *(and so,* $m(A) = m(\{a\})$*).*

Remark 10.3.19 By the above theorem, one can see that in an atom, the entire "information" is concentrated in each of its points.

Theorem 10.3.20 *Suppose* $T = \{t_1, \ldots, t_n\}$ *is a Hausdorff topological space and it is also an atom of a monotone, null-additive regular set function* $m : \mathcal{P}(T) \to \mathbb{R}_+$*. Then* m *is a q-measure.*

Proof Obviously, T is a compact space, so it is locally compact. Also, $\mathcal{B} = \mathcal{P}(T)$.

Since T is an atom, by Theorem 10.3.18 there exists $t_1 \in T$ so that $m(\{t_2, \ldots, t_n\}) = m(T\backslash\{t_1\}) = 0$, whence $m(\{t_2\}) = \ldots = m(\{t_n\}) = 0$.

In consequence, for every $A \subset T$, if $t_1 \notin A$, then $m(A) = 0$ and if $t_1 \in A$, then $m(A) = m(\{t_1\}) = m(T)$.

Now, consider arbitrary pairwise disjoint $A, B, C \in \mathcal{P}(T)$.

If $t_1 \notin A \cup B \cup C$, then $t_1 \notin A, t_1 \notin B, t_1 \notin C$, so the conclusion follows.

If $t_1 \in A \cup B \cup C$, suppose without any lack of generality that $t_1 \in A, t_1 \notin B, t_1 \notin C$. Then $m(B \cup C) = m(B) = m(C) = 0, m(A \cup B \cup C) = m(T) = m(A \cup B) = m(A \cup C) = m(A)$ and the proof finishes.

10.3.2 *Minimal atoms*

In this subsection, let T be an abstract nonvoid set, $\mathcal{C}$ a ring of subsets of T, X a Banach space a monotone set multifunction which satisfies $\mu(\emptyset) = \{0\}$.

Definition 10.3.21 We say that a set $A \in \mathcal{C}$ is:

(i) a minimal atom of μ if $\mu(A) \supsetneq \{0\}$ and for every $B \in \mathcal{C}$, $B \subseteq A$, we have either $\mu(B) = \{0\}$ or $A = B$;

(ii) (Gavriluţ [15–17], Gavriluţ and Croitoru [19–21]) an atom of μ if $\mu(A) \supsetneq \{0\}$ and for every $B \in \mathcal{C}$, $B \subseteq A$, we have either $\mu(B) = \{0\}$ or $\mu(A \setminus B) = \{0\}$;
(iii) (Gavriluţ [15–17], Gavriluţ and Croitoru [19–21]) a pseudo-atom of μ if $\mu(A) \supsetneq \{0\}$ and for every $B \in \mathcal{C}, B \subseteq A$, we have either $\mu(B) = \{0\}$ or $\mu(A) = \mu(B)$.

It is obvious that there exist atoms which are not minimal atoms.

In what follows, suppose, moreover, that μ is monotone with respect to the inclusion of sets.

Remark 10.3.22

(i) Any minimal atom is also an atom (and a pseudo-atom), so, for $\mathcal{MA}$, the collection of all minimal atoms of μ, we have

$$\begin{aligned}\mathcal{MA} = \{A \in \mathcal{C};\ \mu(A) \supsetneq \{0\} &\text{ and for every } B \in \mathcal{C},\ B \subsetneq A, \text{ we have}\\ &\mu(B) = \{0\}\} \subseteq \mathcal{A} = \{A \in \mathcal{C};\ \mu(A) \supsetneq \{0\}\\ &\text{and for every } B \in \mathcal{C},\ B \subseteq A \text{ we have either}\\ &\mu(B) = \{0\} \text{ or } \mu(A \setminus B) = \{0\}\},\end{aligned}$$

where $\mathcal{A}$ is here the collection of all atoms of μ;
(ii) If, moreover, μ is null-additive, then any atom of μ is also a pseudo-atom;
(iii) If A is a minimal atom of μ, then for every $B \subsetneq A$, we have $\mu(B) \supsetneq \{0\}$;
(iv) If $m : \mathcal{C} \to [0, \infty)$ is monotone, $m(\emptyset) = 0$ and $\mu : \mathcal{C} \to \mathcal{P}_f(\mathbb{R})$, $\mu(A) = [0, m(A)]$, for every $A \in \mathcal{C}$, then a set $A \in \mathcal{C}$ is an atom/pseudo-atom/minimal atom of μ if and only if the same is A for m in the sense of Mesiar et al. [40], Ouyang et al. [45].
μ is called the set multifunction induced by the set function m;
(v) If $\mathcal{C} = \mathcal{B}$ and $\mu : \mathcal{B} \to \mathcal{P}_f(X)$, then for every minimal atom A of μ and for every $a \in A$, we have $\mu(A \setminus \{a\}) = \{0\}$. If, moreover, μ is null-additive, then $\mu(A) = \mu(\{a\})$. In a physical interpretation, in this case, information is the same in each point.

In consequence, one can easily construct different examples involving minimal atoms with respect to the set multifunction induced by a set function, starting from the examples given in Mesiar et al. [40], Ouyang et al. [45].

Proposition 10.3.23 *If $\mu : \mathcal{C} \to \mathcal{P}_f(X)$ is null-null-additive and $A, B \in \mathcal{C}$ are two different minimal atoms of μ, then $A \cap B = \emptyset$.*

Proof Suppose that, on the contrary, there exist two non-disjoint, different minimal atoms $A, B \in \mathcal{C}$ of μ. Since $A \setminus (A \cap B) = A \setminus B \subseteq A$ and $A \cap B \subseteq B$, then $[\mu(A \setminus B) = \{0\}$ or $A \setminus B = A]$ and $[\mu(A \cap B) = \{0\}$ or $A \cap B = B]$.

(i) If $\mu(A \setminus B) = \{0\}$, $\mu(A \cap B) = \{0\}$, since μ is null-null-additive, we get that $\mu(A) = \{0\}$, a contradiction.
(ii) If $A \setminus B = A$, then $A \cap B = \emptyset$, a contradiction.

(iii) If $\mu(A \setminus B) = \{0\}$ and $A \cap B = B$, then $B \subseteq A$, so $\mu(B) = \{0\}$ (or $B = A$, a contradiction), so again by the null-null-additivity of μ, we have $\mu(A) = \{0\}$, which is a false.

Evidently, if $A \in \mathcal{C}$ is a minimal atom of μ, then it does not exist another different minimal atom $A_1 \in \mathcal{C}$ of μ so that $A_1 \subset A$.

Proposition 10.3.24

(i) If T is finite, then for every $A \in \mathcal{C}$, with $\mu(A) \supsetneq \{0\}$, there exists $B \in \mathcal{C}$, $B \subseteq A$, which is a minimal atom of μ.
(ii) If, moreover, A is an atom of μ and μ is null-additive, then $\mu(A) = \mu(B)$ and the set B is unique.

Proof

(i) Let be the set $\mathcal{M} = \{M \in \mathcal{C}, M \subseteq A, \mu(M) \supsetneq \{0\}\}$. Obviously, $\mathcal{M} \neq \emptyset$, since $A \in \mathcal{C}$. We observe that any minimal element of $\mathcal{M}$ is a minimal atom of μ. Indeed, let $M \in \mathcal{M}$ be a minimal element of $\mathcal{M}$. Then there cannot exist $D \in \mathcal{M}$ so that $D \subseteq M$ and $D \neq M$ $(*)$.
Since $M \in \mathcal{M}$, then $M \in \mathcal{C}$, $M \subseteq A$, $\mu(M) \supsetneq \{0\}$.
We prove that M is a minimal atom of μ. Indeed, for every $S \subseteq M$, $S \in \mathcal{C}$, we have either $\mu(S) = \{0\}$ or $\mu(S) \supsetneq \{0\}$. In the latter case, we have either $S = M$ or $S \neq M$, which is in contradiction with $(*)$.
(ii) If, on the contrary there are two different minimal atoms B_1 and B_2 of μ, then $\mu(A \setminus B_1) = \mu(A \setminus B_2) = \{0\}$, whence $\mu(A) = \{0\}$, a contradiction.

Proposition 10.3.25 (Self-similarity of Minimal Atoms) *Any subset $B \in \mathcal{C}$, with $\mu(B) \supsetneq \{0\}$ of a minimal atom $A \in \mathcal{C}$ of μ is a minimal atom of μ, too.*

Proof Let $A \in \mathcal{C}$ be a minimal atom of μ and consider any $B \in \mathcal{C}$, with $\mu(B) \supsetneq \{0\}$, $B \subseteq A$. We prove that B is a minimal atom of μ. Indeed, for any $C \in \mathcal{C}$, $C \subseteq B$, then $C \subseteq A$, so either $\mu(C) = \{0\}$ or $C = A$, whence $C = B$.

Example 10.3.26

(i) Suppose that $\mu_1, \mu_2 : \mathcal{C} \to \mathbb{P}_f(\mathbb{R})$ are two monotone set multifunctions such that $\mu_1(\emptyset) = \mu_2(\emptyset) = \{0\}$ and $\mu_1(A) \subseteq \mu_2(A)$, for every $A \in \mathcal{C}$ (for instance, $\mu_1, \mu_2 : \mathcal{C} \to \mathbb{P}_f(\mathbb{R})$, $\mu_1(A) = [0, m_1(A)]$, $\mu_2(A) = [0, m_2(A)]$, for every $A \in \mathcal{C}$, $m_1, m_2 : \mathcal{C} \to \mathbb{R}_+$ being monotone, $m_1(A) \leq m_2(A)$, for every $A \in \mathcal{C}$, $m_1(\emptyset) = m_2(\emptyset) = 0$). Then any minimal atom of μ_2 is a minimal atom of μ_1.
(ii) Let be $\mu : \mathcal{C} \to \mathbb{P}_f(\mathbb{R})$, $\mu(A) = [-m_1(A), m_2(A)]$, for every $A \in \mathcal{C}$, where $m_1, m_2 : \mathcal{C} \to \mathbb{R}_+$, $m_1(\emptyset) = m_2(\emptyset) = 0$. Then a set $A \in \mathcal{C}$ is a minimal atom of μ iff A is a minimal atom for both m_1 and m_2 in the sense of Mesiar et al. [40], Ouyang et al. [45].
(iii) If $\mu : \mathcal{C} \to \mathbb{P}_f(\mathbb{R})$, $\mu(A) = \{m(A)\}$, for every $A \in \mathcal{C}$, where $m : \mathcal{C} \to \mathbb{R}_+$, $m(\emptyset) = 0$, then a set $A \in \mathcal{C}$ is a minimal atom of μ iff A is a minimal atom for m in the sense of Mesiar et al. [40], Ouyang et al. [45].

In this way, one observes that our Definition 10.3.21(i) generalizes to the set-valued case the corresponding notion introduced by Mesiar et al. [40], Ouyang et al. [45].

Definition 10.3.27

(i) If $\mu : \mathcal{C} \to \mathbb{P}_f(X)$, we consider the variation of μ, $\overline{\mu} : \mathcal{P}(T) \to [0, \infty]$, defined for every $A \in \mathcal{P}(T)$ by:

$$\overline{\mu}(A) = \sup\left\{ \sum_{i=1}^{p} |\mu(A_i)|;\ A = \bigcup_{i=1}^{p} A_i,\ A_i \in \mathcal{C}, \forall i = \overline{1,p},\ A_i \cap A_j = \emptyset,\ i \neq j \right\}.$$

(ii) We say that μ is of finite variation if $\overline{\mu}(T) < \infty$.

Remark 10.3.28 For every $A \in \mathcal{C}$, we have $\overline{\mu}(A) \geq |\mu(A)|$. In consequence, if $A \in \mathcal{C}$ is a minimal atom of $\overline{\mu}$ (in the sense of Mesiar et al. [40], Ouyang et al. [45]), then A is a minimal atom of μ. Moreover, conversely, if $A \in \mathcal{C}$ is a minimal atom of μ, then it is also an atom of μ, so $\overline{\mu}(A) \geq |\mu(A)|$, whence A is a minimal atom of $\overline{\mu}$.

Remark 10.3.29

(i) Any set $A \in \mathcal{C}$ that can be written as $\bigcup_{i=1}^{p} A_i$ (where for every $i = \overline{1,p}$, $A_i \in \mathcal{C}$ are different minimal atoms of μ), is partitioned in fact in this way, since by Proposition 10.3.23 we have $A_i \cap A_j = \emptyset$, $i \neq j$.

Since any minimal atom is an atom, then in this case $\overline{\mu}(A_i) = |\mu(A_i)|$, for every $i = \overline{1,p}$. In consequence, if, moreover, μ is a multisubmeasure (in the sense of Gavriluţ [15]) of finite variation, then by [15], $\overline{\mu}$ is finitely additive, so $\overline{\mu}(A) = \sum_{i=1}^{p} |\mu(A_i)|$.

(ii) (non-decomposability of minimal atoms) Any minimal atom $A \in \mathcal{C}$ cannot be partitioned (its only partition is $\{A, \emptyset\}$).

The converse of the last statement also holds:

Proposition 10.3.30 *Any non-partitionable atom $A \in \mathcal{C}$ is a minimal atom.*

Proof Since A is an atom, then $\mu(A) \supsetneq \{0\}$. On the other hand, because A is non-partitionable, there cannot exist two nonvoid disjoint subsets of A, let us say $A_1, A_2 \in \mathcal{C}$.

Let now be any $B \in \mathcal{C}$, with $B \subseteq A$. One has either $\mu(B) = \{0\}$ (which is fine) or $\mu(B) \supsetneq \{0\}$. In the latter case, the only possibility is $B = A$ (if not, $\{A \setminus B, B\}$ is a partition of A, which is false).

Corollary 10.3.31 *An atom is minimal if and only if it is not partitionable.*

Theorem 10.3.32 *If $\mathcal{C} = \mathcal{B}$ and if μ is null-additive and regular, then any minimal atom $A \in \mathcal{B}$ of μ is a singleton.*

Proof There $\exists! a \in A$ so that $\mu(A) = \mu(\{a\})$. Then either $A = \{a\}$, or $\mu(\{a\} = \{0\}$, in which case $\mu(A) = \{0\}$, which is absurd, since A is an atom.

Theorem 10.3.33 *If T is finite, μ is null-additive and $\{A_i\}_{i=\overline{1,p}}$ is the set of all minimal different atoms contained in a set $A \in \mathcal{C}$, with $\mu(A) \supsetneq \{0\}$ (this set exists by Proposition 10.3.24), then $\mu(A) = \mu(\bigcup\limits_{i=1}^{p} A_i)$ (so, the minimal atoms are the only ones which are important from the "measurement" viewpoint).*

Proof Obviously, $\mu(A \setminus \bigcup\limits_{i=1}^{p} A_i) = \{0\}$ (if not, by Proposition 10.3.24(i), there exists another minimal atom of μ). Consequently, by the null-additivity of μ, $\mu(A) = \mu(\bigcup\limits_{i=1}^{p} A_i)$.

Corollary 10.3.34 *If T is finite, μ is a multisubmeasure and $\{A_i\}_{i=\overline{1,p}}$ is the set of all minimal different atoms contained in a set $A \in \mathcal{C}$, with $\mu(A) \supsetneq \{0\}$, then $\mu(A) \subseteq \overline{\sum\limits_{i=1}^{p} \mu(A_i)}$ (where the symbol "$\overline{\subseteq}$" denotes the closure with respect to the topology induced by the norm of X). Moreover, $|\mu(A)| \leq \sum\limits_{i=1}^{p} |\mu(A_i)|$.*

Proof The statement is immediate by the above theorem since μ is in particular null-additive and $|\mu|$ is a submeasure (in the sense of Drewnowski [13]).

In consequence, although $\bigcup\limits_{i=1}^{p} A_i \subseteq A$, we have $\mu(\bigcup\limits_{i=1}^{p} A_i) = \mu(A) \subseteq \overline{\sum\limits_{i=1}^{p} \mu(A_i)}$.

10.3.3 Coherence and decoherence through Young type experiments

In Quantum Mechanics, when a wavefunction becomes coupled to its environment, the objects involved interacting with the surroundings, the decoherence phenomenon occurs. It is also known as the"wavefunction collapse" and it allows the classical limit to emerge on the macroscopic scale from a set of quantum events. After decoherence has occurred, the system's components can no longer interfere, so one could assign a well-defined probability to each possible decoherent outcome.

In light's classical theory, the intensity of the light in an arbitrary point is determined by the square amplitude of the light. For instance, in Young's two-slit experiment, the intensity of the light on the detector screen is given by the square amplitude of the wave obtained through the overlapping (superposition) of the secondary waves originating from each slit. Of course, this classical wave theory cannot be used in this case since it ignores the corpuscular character of the light. However, by analogy, it suggests that in Quantum Mechanics, it can be introduced either a wavefunction which satisfies the Schrödinger equation. The wavefunction is a complex quantity, while the states density is a real one. We expect then, that the states density $\rho(x, y, z, t)$ to find the particle in a given point from the volume V, in a vicinity of the point of coordinates (x, y, z, t) at a momentum t should be proportional with $|\Psi|^2$, that is

$$\rho(x, y, z, t) \equiv |\Psi(x, y, z, t)|^2$$

Let Ψ_1 be the wavefunction in a given point from the screen where the interference field is localized, corresponding to the waves propagated through the slit 1. Similarly, let Ψ_2 be the wavefunction in the same point, corresponding to the waves propagated through the slit 2. The two intensity distributions, corresponding to the "experiments" performed with only one open slit are determined by the respective states densities (probability distributions)

$$\rho_1 \equiv |\Psi_1|^2, \rho_2 \equiv |\Psi_2|^2.$$

On the other hand, when both slits are open, the wavefunction is given by the sum of the two contributions Ψ_1 and Ψ_2:

$$\Psi \equiv \Psi_1 + \Psi_2.$$

The corresponding states density (probability distribution)

$$\rho \equiv |\Psi_1 + \Psi_2|^2$$

determines then the intensity of the "structure" from the interference field.

Let us explain in the following, Ψ_1 and Ψ_2 in the form

$$\Psi_1 = \sqrt{\rho_1} e^{i\theta_1}, \Psi_2 = \sqrt{\rho_2} e^{i\theta_2}.$$

It results

$$\rho \equiv |\Psi|^2 = |\Psi_1|^2 + |\Psi_2|^2 + 2\Re e\{|\Psi_1| \cdot |\Psi_2| \exp[i(\theta_2 - \theta_1)]\} \equiv$$
$$\rho_1 + \rho_2 + 2\sqrt{\rho_1 \rho_2} \cos \Delta\theta, \Delta\theta = \theta_2 - \theta_1.$$

Now, if the term $\cos \Delta\theta$ is a time functional

$$\cos \Delta\theta \equiv \cos \Delta\theta(t)$$

then the system is decoherent (the interference field does not exist). If

$$\cos \Delta\theta = \text{const.}$$

then the system is coherent (there exists an interference field).

Remark 10.3.35 Decoherence is a precise formulation of the basic principle underlying the Schrödinger's cat thought experiment—the outcome of a quantum event remains undetermined until the system interacts with its environment. Thus, using the decoherence function, one could define the probabilities of all decoherent outcomes for a particular event by quantifying the amount of interference among system's various components. So, interference has an important role in the mathematical formulation of quantum mechanics.

One can define functions related to interference that can be used in order to obtain q-measures:

Definition 10.3.36 (Gavriluţ and Agop [18]) Suppose T is an abstract space and $\mathcal{A}$ is an algebra of subsets of T. A function $D : \mathcal{A} \times \mathcal{A} \to \mathbb{C}$, the set of all complex numbers is said to be a decoherence function if the following conditions hold:

(i) $D(A, B) = \overline{D(B, A)}$, for every $A, B \in \mathcal{A}$ (where the symbol "‾" denotes here the complex conjugate);
(ii) $D(A, A) \geq 0$, for every $A \in \mathcal{A}$;
(iii) $|D(A, B)| \leq D(A, A) \cdot D(B, B)$, for every $A, B \in \mathcal{A}$;
(iv) $D(A \cup B, C) = D(A, C) + D(B, C)$, for every disjoint $A, B \in \mathcal{A}$ and every $C \in \mathcal{A}$.

Remark 10.3.37 (Gavriluţ and Agop [18])

(i) Since $D(A, A) \in \mathbb{R}$, the conditions (ii) and (iii) in Definition 10.3.36 are justified.
(ii) By (i), for arbitrary $A, B \in \mathcal{A}$ representing quantum objects, $Re[D(A, B)]$ can be interpreted as the interference between A and B, as we remark in what follows:

Proposition 10.3.38 (Gavriluţ and Agop [18]) *If $D : \mathcal{A} \times \mathcal{A} \to \mathbb{C}$ is a decoherence function, then $m : \mathcal{A} \to \mathbb{C}$, $m(A) = D(A, A)$ is a q-measure.*

Example 10.3.39 If V is a pre-Hilbert space and if $m : \mathcal{A} \to V$ is finitely additive, then $D : \mathcal{A} \times \mathcal{A} \to \mathbb{C}$,

$$D(A, B) =< m(A), m(B) >,$$

for every $A, B \in \mathcal{A}$ is a decoherence function. Particularly, if $m : \mathcal{A} \to \mathbb{C}$ is finitely additive (often interpreted as a quantum amplitude), then one can define the decoherence function defined for every $A, B \in \mathcal{A}$ by

$$D(A, B) = m(A) \cdot \overline{m(B)}.$$

The corresponding q-measure is $\widetilde{m} : \mathcal{A} \to \mathbb{C}$,

$$\widetilde{m}(A) = D(A, A) = m(A) \cdot \overline{m(A)} = |m(A)|^2,$$

for every $A \in \mathcal{A}$ ("$\overline{m(A)}$" indicates the complex conjugate and "$|\cdot|$", the modulus of a complex number).

Remark 10.3.40

(i) (Gavriluţ and Agop [18]) If $A, B \in \mathcal{A}$ are disjoint, then $\widetilde{m}$ from Example 10.3.39 is not grade-1-additive. Indeed,

$$\begin{aligned}\widetilde{m}(A \cup B) &= |m(A \cup B)|^2 = |m(A) + m(B)|^2 = \\ &= |m(A)|^2 + |m(B)|^2 + 2\Re\mathrm{e}[m(A)\overline{m(B)}] = \\ &= \widetilde{m}(A) + \widetilde{m}(B) + 2\Re\mathrm{e} D(A, B).\end{aligned}$$

Also, $\widetilde{m}(A \cup B) = \widetilde{m}(A) + \widetilde{m}(B)$ iff $\Re\mathrm{e} D(A, B) = 0$, i.e., interference is represented by the real part of a decoherence function.

(ii) If $m : \mathcal{A} \to \mathbb{R}$ is a real valued submeasure of finite variation, then

$$D : \mathcal{A} \times \mathcal{A} \to \mathbb{R},\ D(A, B) =< \overline{m}(A), \overline{m}(B) >$$

is a decoherence function, where in this case $\overline{m}$ is the variation of m.

(iii) If X is a Banach space and $\mu : \mathcal{A} \to \mathcal{P}_f(X)$ is a multisubmeasure of finite variation, then

$$D : \mathcal{A} \times \mathcal{A} \to \mathbb{R},\ D(A, B) =< \overline{\mu}(A), \overline{\mu}(B) >$$

is a decoherence function, where $\overline{\mu}$ denotes the variation of μ.

If $A, B \in \mathcal{A}$ are atoms of μ, then, moreover,

$$D(A, B) =< \overline{\mu}(A), \overline{\mu}(B) >=< |\mu(A)|, |\mu(B)| > .$$

10.4 Fractal Mechanics and some applications

Let us admit, accordingly to the above sections, that the structural units of any complex system are moving on continuous and non-differentiable curves (fractal curves). Then the dynamics of any complex systems are functional by means of the scale covariance principle (Mercheş and Agop [39], Nottale [44]): the physics laws which describe the dynamics of the complex systems are invariant with respect to scale transformations (Mazilu and Agop [38]).

10.4.1 Fractal operator and its implications

Theorem 10.4.1 *If the scale covariance principle is functional, then the transition from the dynamics of the classical physics (differential physics) to the dynamics of the fractal physics (non-differentiable physics) can be implemented by replacing the usual derivative operator d/dt by the fractal operator $\hat{d}/dt$*

$$\frac{\hat{d}}{dt} = \partial_t + \hat{V}^l \partial_l - i\lambda(dt)^{(2/D_F)-1}\partial_l \partial^l \tag{10.1}$$

where

$$\partial_t = \frac{\partial}{\partial t}, \partial_l = \frac{\partial}{\partial X^l}, \partial_l \partial^l = \frac{\partial}{\partial X_l}(\frac{\partial}{\partial X^l}) \tag{10.2}$$

$$\hat{V}^l = V^l_D - iV^l_F, i = \sqrt{-1}, l = 1, 2, 3 \tag{10.3}$$

In the above relations, X^l are the fractal spatial coordinates, t is the non-fractal temporal coordinate with the role of an affine parameter, $\hat{V}^l$ is the velocities complex field, V^l_D is the real part of the complex velocity which is independent on the scale resolution dt, V^l_F is the imaginary part of the complex velocity which is dependent on the scale resolution, λ is the diffusion coefficient associated to the fractal-non-fractal transition and D_F is the fractal dimension of the motion curve.

For D_F one can choose different definitions, as fractal dimension in the sense of Kolmogorov, fractal dimension in the sense of Hausdorff-Besikovici, etc. (Mercheş and Agop [38], Nottale [44]).

Once chosen the definition of the fractal dimension, it has to remain constant during the whole analysis of the complex systems dynamics.

Proof The proof of the above statements is given in Mercheş and Agop [39].

The fractal operator (10.1) plays the role of the scale covariant derivative, namely it is used to write the fundamental equations of complex systems dynamics in the same form as in the classical (differentiable) case.

Theorem 10.4.2 *Applying the operator (10.1) to the complex velocity field (10.3), the equation of motion (geodesics equation), in the absence of an external scalar potential U takes the form:*

$$\frac{\hat{d}\hat{V}^i}{dt} = \partial_t \hat{V}^i + \hat{V}^l \partial_l \hat{V}^i - i\lambda(dt)^{(2/D_F)-1}\partial_l \partial^l \hat{V}^i \equiv 0, \tag{10.4}$$

while in the presence of an external scalar potential U the equation of motion (geodesics equation) becomes:

$$\frac{\hat{d}\hat{V}^i}{dt} = \partial_t \hat{V}^i + \hat{V}^l \partial_l \hat{V}^i - i\lambda(dt)^{(2/D_F)-1}\partial_l \partial^l \hat{V}^i = -\partial^i U. \tag{10.5}$$

Proof This result can be directly verified.

In the case of irrotational motions of the complex system structural units, the complex velocity field (10.3) satisfies the restriction:

$$\varepsilon_{ilk}\partial^l \hat{V}^k = 0 \tag{10.6}$$

with ε_{ilk} the Levi-Civita pseudo-tensor.

From here, $\hat{V}^i$ is expressed through the gradient of the complex scalar function $\ln \Psi$, called the scalar potential of the complex velocities field, in the form

$$\hat{V}^i = -2i\lambda(dt)^{(2/D_F)-1}\partial^i \ln \Psi \tag{10.7}$$

Corollary 10.4.3 *A Schrödinger type fractal equation, which is free of any "external constraints":*

$$\lambda^2(dt)^{(4/D_F)-2}\partial^l \partial_l \Psi + i\lambda(dt)^{(2/D_F)-1}\partial_t \Psi \equiv 0 \tag{10.8}$$

or a Schrödinger type fractal equation in the presence of an "external constraint" U, respectively:

$$\lambda^2(dt)^{(4/D_F)-2}\partial^l \partial_l \Psi + i\lambda(dt)^{(2/D_F)-1}\partial_t \Psi - \frac{U}{2}\Psi \equiv 0 \tag{10.9}$$

can be obtained substituting the relation (10.7), either in the geodesics equation (10.4), or in the motion equation (10.5).

Proof This result can be directly obtained following the procedure from Mercheş and Agop [39].

Remark 10.4.4 Assuming now that the motions of the structural units of any complex system take place on Peano curves (i.e., for $D_F = 2$), at Compton scale resolution (i.e., for $\lambda = \hbar/2m_0$, where $\hbar$ is the reduced Planck constant and m_0 the rest mass of the structural unit), Equations (10.8) and (10.9) reduce to the standard Schrödinger equations:

$$\frac{h^2}{2m_0}\Delta\Psi + i\hbar\partial_t\Psi = 0 \tag{10.10}$$

and

$$\frac{h^2}{2m_0}\Delta\Psi + i\hbar\partial_t\Psi - U\Psi = 0 \tag{10.11}$$

respectively.

Practically, we discuss here about fractality through Markov type stochastic processes.

Remark 10.4.5 Since by means of this mathematical procedure, Quantum Mechanics becomes a particular case of Fractal Mechanics (in the fractal dimension $D_F = 2$ at Compton scale resolution $\lambda = \hbar/2m_0$), all the results of Quantum Mechanics can be extended (generalized) to any resolution scale. Thus, "fundamental concepts" of Quantum Mechanics as quantum entanglement, quantum superposition, quantum information, etc. have to be substituted by those of "fractal entanglement", "fractal superposition", "fractal information" (Agop et al. [2–5], Audenaert [6], Grigorovici et al. [24]), etc. That is why, in Section 10.5 we shall substitute the notion of an atom/pseudo-atom with that of a fractal atom/fractal pseudo-atom, respectively. On the properties of these new mathematical objects we shall discuss in a future work.

10.4.2 Stationary dynamics of a complex system structural units in the fractal Schrödinger representation

Theorem 10.4.6 *In the case of the stationary dynamics of a complex system, the "synchronization" of its structural units implies "hidden symmetries". The explicitation of these "hidden symmetries" can be achieved through a continuous group with three parameters (the homographic group).*

Proof The proof of the theorem is based on the properties of the solutions of the Schrödinger type fractal equation for the stationary one-dimensional case.

For this reason, the Schrödinger type fractal equation (10.8) in the one-dimensional case becomes:

$$\lambda^2 (dt)^{(4/D_F)-2} \partial_{xx} \Psi(x,t) + i\lambda (dt)^{(2/D_F)-1} \partial_t \Psi = 0 \tag{10.12}$$

By means of the solution

$$\Psi(x,t) = \theta(x) \exp[-\frac{i}{2m_0 \lambda (dt)^{(2/D_F)-1}} Et] \tag{10.13}$$

where E is the energy of a complex system structural unit, Equation (10.12) becomes:

$$\partial_{xx}\theta(x) + k_0^2 \theta(x) = 0 \tag{10.14}$$

$$k_0^2 = \frac{E}{2m_0 \lambda (dt)^{(4/D_F)-2}} \tag{10.15}$$

i.e., a stationary Schrödinger type fractal equation.

The most general solution of Equation (10.14) can be written in the form:

$$\theta(x) = he^{i(k_0x+\varphi)} + \overline{h}e^{-i(k_0x+\varphi)} \tag{10.16}$$

with h a complex amplitude, $\overline{h}$ its complex conjugate and φ a phase.

This solution describes a complex system structural units of the same "characteristic" k_0, in which the structural unit is identified by means of the parameters $h, \overline{h}$ and $k = e^{i\varphi}$.

Now, a question arises. Which is the relation among the structural units of the complex system having the same k_0? The mathematical answer to this question can be obtained if we admit that all we intend here is to find a way to switch from a triplet of numbers—the initial conditions—of a structural unit, to the same triplet of another structural unit having the same k_0.

This passage implies a "hidden symmetry" which is explicited in the form of a continuous group with three parameters, group that is simple transitive and which can be constructed using a certain definition of k_0.

We start from the idea that the ratio between two fundamental solutions of Equation (10.16) is a solution of Schwartz's nonlinear equation (Mihăileanu [42]):

$$\{\tau_0(x), x\} = 2k_0^2,\ \tau_0(x) = e^{-2ik_0x} \tag{10.17}$$

where the curly brackets define Schwartz's derivative of τ_0 with respect to x,

$$\{\tau_0(x), x\} = \partial_x\left(\frac{\partial_{xx}\tau_0}{\partial_x\tau_0}\right) - \frac{1}{2}\left(\frac{\partial_{xx}\tau_0}{\partial_x\tau_0}\right)^2. \tag{10.18}$$

This equation proves to be a veritable definition of k_0, as a general characteristic of a complex system of structural units which can be swept through a continuous group with three parameters—the homographic group.

Indeed, Equation (10.17) is invariant with respect to the dependent variable change:

$$\tau(x) = \frac{a\tau_0(x) + b}{c\tau_0(x) + d},\ a, b, c, d \in \mathbb{R} \tag{10.19}$$

and this statement can be directly verified.

In this way, $\tau(x)$ characterizes another structural unit of the same k_0, which allows us to state that, starting from a standard structural unit, we can sweep the entire complex system of structural units having the same k_0, when we are not conditioning (we leave it free) the three ratios $a : b : c : d$ in Equation (10.19).

We can make even more accurate the correspondence between a homographic transformation and a structural unit of the complex system, by associating to every structural unit of the complex system, a "personal" $\tau(x)$ by the relation:

$$\tau_1(x) = \frac{h + \overline{h}k\tau_0(x)}{1 + k\tau_0(x)},\ k = e^{-2i\varphi}. \tag{10.20}$$

Let us observe that τ_0 and τ_1 can be used freely one in place of another and this leads us to the following transformation group for the initial conditions:

$$h \leftrightarrow \frac{ah+b}{ch+d}, \bar{h} \leftrightarrow \frac{a\bar{h}+b}{c\bar{h}+d}, k \leftrightarrow \frac{c\bar{h}+d}{ch+d}k. \tag{10.21}$$

This group is simple transitive: to a given set of values $(a/c, b/c, d/c)$ will correspond a single transformation and only one of the group.

Remark 10.4.7 The group (10.21) works as a group of "synchronization" among the various structural units of the complex system, process to which the amplitudes and phases of each of them obviously participate, in the sense that they are correlated, too. More precisely, by means of (10.21), the phase of k is only moved with a quantity depending on the amplitude of the structural unit of complex system at the transition among various structural units of the complex system. But not only that, the amplitude of the structural unit of the complex system is also affected homographically.

The usual "synchronization" manifested through the delay of the amplitudes and phases of the structural units of the complex system must represent here only a totally particular case.

Theorem 10.4.8 *In the "field variables" space of the synchronization group one can a priori build a probabilistic theory based on its elementary measure, as an elementary probability. Then the invariant function of the synchronization group becomes the repartition density of an elementary probability.*

Proof The proof of these statements is based on the differential and integral properties of the homographic group. Thus, considering a specific parametrization of the group (10.21), the infinitesimal generators (Mercheş and Agop [39]):

$$\hat{B}_1 = \frac{\partial}{\partial h} + \frac{\partial}{\partial \bar{h}}, \hat{B}_2 = h\frac{\partial}{\partial h} + \bar{h}\frac{\partial}{\partial \bar{h}}, \hat{B}_3 = h^2\frac{\partial}{\partial h} + \bar{h}^2\frac{\partial}{\partial \bar{h}} + (h-\bar{h})k\frac{\partial}{\partial k} \tag{10.22}$$

satisfy the commutation relations:

$$[\hat{B}_1, \hat{B}_2] = \hat{B}_1, [\hat{B}_2, \hat{B}_3] = \hat{B}_3, [\hat{B}_3, \hat{B}_1] = -2\hat{B}_2. \tag{10.23}$$

The structure of the group (10.21) is given by Equation (10.23) so that the only non-zero structure constants should be:

$$C_{12}^1 = C_{23}^3 = -1, C_{31}^2 = -2 \tag{10.24}$$

Therefore, the invariant quadratic from is given by the "quadratic" tensor of the group (10.21):

$$C_{\alpha\beta} = C_{\alpha\nu}^{\mu} C_{\beta\mu}^{\nu} \tag{10.25}$$

where summation over repeated indices is understood. Using (10.24) and (10.25), the tensor $C_{\alpha\beta}$ writes:

$$C_{\alpha\beta} = \begin{pmatrix} 0 & 0 & -4 \\ 0 & 2 & 0 \\ -4 & 0 & 0 \end{pmatrix} \tag{10.26}$$

meaning that the invariant metric of the group (10.21) has the form:

$$\frac{ds^2}{g^2} = \omega_0^2 - 4\omega_1\omega_2 \tag{10.27}$$

with g an arbitrary factor and $\omega_\alpha, \alpha = 0, 1, 2$ three differential 1-forms (Flanders [14]), absolutely invariant through the group (10.21). Barbilian takes these 1-forms as being given by the relations (Barbilian [7], Mercheş and Agop [39]):

$$\omega_0 = -i(\frac{dk}{k} - \frac{dh + d\overline{h}}{h - \overline{h}}), \omega_1 = \frac{dh}{(h - \overline{h})k}, \omega_2 = \frac{-kd\overline{h}}{(h - \overline{h})} \tag{10.28}$$

so that the metric (10.27) becomes

$$\frac{ds^2}{g^2} = -(\frac{dk}{k} - \frac{dh + d\overline{h}}{h - \overline{h}})^2 + 4\frac{dhd\overline{h}}{(h - \overline{h})^2}. \tag{10.29}$$

It is worthwhile to mention a property connected to the integral geometry: the group (10.21) is measurable. Indeed, it is simply transitive and, since its structure vector

$$C_\alpha = C^{\nu}_{\nu\alpha} \tag{10.30}$$

is identically null, as it can be seen form (10.24), this means that it possess the invariant function

$$F(h, \overline{h}, k) = -\frac{1}{(h - \overline{h})^2 k} \tag{10.31}$$

which is the inverse of the modulus of determinant of a linear system obtained on the basis of infinitesimal transformations of the group (10.21).

As a result, in the space of the field variables $(h, \overline{h}, k)$ one can a priori construct a probabilistic theory in the sense of Jaynes [33] (on the circumstances left unspecified in an experiment), based on the elementary measure of the group (10.21),

$$dP(h, \overline{h}, k) = -\frac{dh \Lambda d\overline{h} \Lambda dk}{(h - \overline{h})^2 k} \tag{10.32}$$

as elementary probability, where Λ denotes the external product of the 1-forms. In such context, the invariant function of the group (10.21), i.e., relation (10.31), becomes the repartition density of the elementary probability (10.32).

Remark 10.4.9 The above results can be re-written in real terms based on the transformation:

$$(h, \overline{h}, k) \to (u, v, \phi) \tag{10.33}$$

which can be made explicit through the relations

$$h = u + iv, \overline{h} = u - iv, k = e^{i\phi} \tag{10.34}$$

Thus, both the operators (10.22) and the 1-forms (10.28) have the expressions:

$$\hat{M}_1 = \frac{\partial}{\partial u}, \hat{M}_2 = u\frac{\partial}{\partial u} + v\frac{\partial}{\partial v}, \hat{M}_3 = (u^2 - v^2)\frac{\partial}{\partial u} + 2uv\frac{\partial}{\partial v} + 2v\frac{\partial}{\partial \phi} \tag{10.35}$$

respectively

$$\Omega^1 \equiv \omega^0 = d\phi + \frac{du}{v}, \Omega^2 = \omega^1 = \cos\phi\frac{du}{v} + \sin\phi\frac{dv}{v}, \Omega^3 = \omega^2 = -\sin\phi\frac{du}{v} + \cos\phi\frac{dv}{v} \tag{10.36}$$

while the 2-form (10.29) reduces to the two-dimensional Lorentz metric

$$-(\Omega^1)^2 + (\Omega^2)^2 + (\Omega^3)^2 = -(d\phi + \frac{du}{v})^2 + \frac{du^2 + dv^2}{v^2} \tag{10.37}$$

Remark 10.4.10 An attitude toward Quantum Mechanics which is suitable for Quantum Gravity in general, and for its application to cosmology in particular, is not so easy to find. A philosophically realistic attitude toward Quantum Mechanics would seem to be more effective than one based on operators which must find their physical meaning in terms of measurements.Where quantum theory differs from classical mechanics (in this view) is in its dynamics, which of course is stochastic rather than deterministic. As such, the theory functions by furnishing probabilities for sets of histories. What ordinarily makes it difficult to regard Quantum Mechanics as in essence a modified form of probability theory, is the peculiar fact that it works with complex amplitudes rather than directly with probabilities, the former being more like square roots of the latter. In this context the above-mentioned whole arsenal of Quantum Mechanics can be extended to fractal manifolds by means of a Jaynes type procedure [33].

Theorem 10.4.11 *The existence of a transport of directions in the Levi-Civita sense in the field variables space substitutes the homographic group with that of spin as a synchronization group.*

Proof Let us focus on the metric (10.29) or (10.37). It is reduced to the metric of Lobachewski's plane in Poincaré representation:

$$\frac{ds^2}{g^2} = 4\frac{dhd\overline{h}}{(h-\overline{h})^2} \tag{10.38}$$

for the condition $\omega_0 = 0$, i.e., in real terms (10.34)

$$d\phi = -\frac{du}{v} \tag{10.39}$$

Since by this restriction the metric (10.37) in the variables (10.34) reduces to Lobachewski's one in Beltrami's representation:

$$\frac{ds^2}{g^2} = -\frac{du^2 + dv^2}{v^2} \tag{10.40}$$

the condition (10.39) defines a parallel transport of vectors in the sense of Levi-Civita (the definition of the parallelism angle in the Lobachewski plane, that is, the form of connection—Agop et al. [1], Mercheş and Agop [39], Grigorovici et al. [24]): the application point of the vector moves on the geodesic, the vector always making a constant angle with the tangent to the geodesic in the current point. Indeed, taking advantage of the fact that the metric of the plane is conformal Euclidean, we can calculate the angle between the initial vector and the vector transported through parallelism, as the integral of the equation (Agop et al. [1], Mercheş and Agop [39]).

$$d\phi = \frac{1}{2}[\frac{\partial}{\partial v}(\ln F)du - \frac{\partial}{\partial u}(\ln F)dv],\ F(u,v) = \frac{1}{v^2} \tag{10.41}$$

along the transport curve.

Since $F(u, v)$ represents the conformal factor of the given metric, introducing it in (10.41), we find (10.39).

Remark 10.4.12 The "ensemble" of the initial conditions of the structural units of the complex system corresponding to the same k_0 can be organized as a geometry of the hyperbolic plane. More precisely, these structural units of the complex system correspond to a situation where their initial conditions can be chosen from among points of a hyperbolic plane.

Remark 10.4.13 The existence of the parallel transport in the sense of Levi-Civita (10.39) implies either the substitution of the operators (10.22) with the operators:

$$\hat{B}_1' = \frac{\partial}{\partial h} + \frac{\partial}{\partial \bar{h}}, \hat{B}_2' = h\frac{\partial}{\partial h} + \bar{h}\frac{\partial}{\partial \bar{h}}, \hat{B}_3' = h^2\frac{\partial}{\partial h} + \bar{h}^2\frac{\partial}{\partial \bar{h}} \tag{10.42}$$

in the case of the representation in complex variables, or the substitution of the operators (10.35) with the operators:

$$\hat{M}_1' = \frac{\partial}{\partial u}, \hat{M}_2' = u\frac{\partial}{\partial u} + v\frac{\partial}{\partial v}, \hat{M}_3' = (u^2 - v^2)\frac{\partial}{\partial u} + 2uv\frac{\partial}{\partial v} \tag{10.43}$$

in the case of the representation in real variables.

Theorem 10.4.14 *Through the correlation phase-amplitude given by the relation (10.39), the operators (10.43) reduce to the spin operators in the null vectors space*

$$\hat{S}_1 = \cos\psi v\frac{\partial}{\partial v} - \sin\psi\frac{\partial}{\partial \psi}, \hat{S}_2 = \sin\psi v\frac{\partial}{\partial v} + \cos\psi\frac{\partial}{\partial \psi}, \hat{S}_3 = i\frac{\partial}{\partial \psi}. \tag{10.44}$$

Precisely, we discuss about the compactification of the angular momentum in the null vectors space in the form of the spin.

Proof The proof of the above statements is given in Grigorovici et al. [24].

Remark 10.4.15 These operators multiplied with the factor $\lambda(dt)^{(2/D_F)-1}$, are identical, with the fractal angular momentum operators Grigorovici et al. [24] in the representations

$$x = v\sin\psi, y = -v\cos\psi, z = iv. \tag{10.45}$$

One can directly verify that, ignoring the factor $\lambda(dt)^{(2/D_F)-1}$, the operators (10.44) are just the fractal spin operators satisfying the same commutation relations as Pauli matrix σ_i $(i = 1, 2, 3)$. They can be interpreted as fractal angular momentum operators in the fractal space of null radius

$$x^2 + y^2 + z^2 = 0. \tag{10.46}$$

Remark 10.4.16

(i) The corresponding variables (v, ψ) are not concrete variables but just only internal freedom degrees.
(ii) By means of condition (10.39), we can select the minimal fractal atoms from the fractal atoms.

10.5 From the standard mathematical atom to the fractal atom by means of a physical procedure

Let T be an abstract nonvoid set, $\mathcal{C}$ a lattice of subsets of T and $m : \mathcal{C} \to \mathbb{R}_+$ an arbitrary set function with $m(\emptyset) = 0$. Evidently, one can immediately generalize the notions of a pseudo-atom/minimal atom, respectively, to this context when $\mathcal{C}$ is only a lattice and not necessarily a ring.

Example 10.5.1

(i) It T is a metric space, then the Hausdorff dimension $\dim_{Haus} : \mathcal{P}(T) \to \mathbb{R}$ (Mandelbrot [37]) is a monotone real function. Evidently, $\dim_{Haus}(\emptyset) = 0$.
(ii) For every $d \geq 0$, the Hausdorff measure $H^d : \mathcal{P}(T) \to \mathbb{R}$ is an outer measure, so, particularly, it is a submeasure.

Remark 10.5.2

(i) The union of two sets A and B having the fractal dimensions D_A, respectively, D_B has the fractal dimension $D_{A\cup B} = \max\{D_A, D_B\}$;
(ii) The intersection of two sets A and B having the fractal dimensions D_A, respectively, D_B has the fractal dimension $D_{A\cap B} = D_A + D_B - d$, where d is the embedding Euclidean dimension (Iannaccone and Khokha [32]).

The following definition is then consistent:

Definition 10.5.3 A pseudo-atom/minimal atom, respectively, $A \in \mathcal{C}$ of m having the fractal dimension D_A is said to be a *fractal pseudo-atom/fractal minimal atom*, respectively.

By Proposition 10.3.11(iii) and Remark 10.5.2, we get:

Proposition 10.5.4 *If $A, B \in \mathcal{C}$ are fractal pseudo-atoms of m and if $m(A \cap B) > 0$, then $A \cap B$ is a fractal pseudo-atom of m and $m(A \cap B) = m(A) = m(B)$.*

References

1. Agahi, H., Mesiar, R., Babakhani, A.: Generalized expectation with general kernel on g-semirings and its applications. Rev. Real Acad. Cienc. Exact. Fís. Natur. Ser. A Mat. **111**(3), 863–875 (2017)
2. Agop, M., Gavriluţ, A., Crumpei, G., Doroftei, B.: Informational non-differentiable entropy and uncertainty. Relations in complex systems. Entropy **16**, 6042–6058 (2014)
3. Agop, M., Gavriluţ, A., Rezuş, E.: Implications of Onicescu's informational energy in some fundamental physical models. Int. J. Mod. Phys. B **29**, 1550045, 19 pp. (2015). https://doi.org/10.1142/S0217979215500459
4. Agop, M., Gavriluţ, A., Ştefan, G., Doroftei, B.: Implications of non-differentiable entropy on a space-time manifold. Entropy **17**, 2184–2197 (2015)

5. Agop, M., Gavriluţ, A., Păun, V.P., Filipeanu, D., Luca, F.A., Grecea, C., Topliceanu, L.: Fractal information by means of harmonic mappings and some physical implications. Entropy **18**, 160 (2016)
6. Audenaert, K.M.R.: Subadditivity of q-entropies for $q > 1$. J. Math. Phys. **48**, 083507 (2007)
7. Barbilian, D.: Die von einer Quantik induzierte Riemannsche Metrik. C. R. Acad. Roum. Sci. **2**, 198 (1937)
8. Boonsri, N., Saejung, S.: Some fixed point theorems for multivalued mappings in metric spaces. Rev. Real Acad. Cienc. Exact. Fís. Natur. Ser. A Mat. **111**(2), 489–497 (2017)
9. Cavaliere, P., Ventriglia, F.: On nonatomicity for non-additive functions. J. Math. Anal. Appl. **415**(1), 358–372 (2014)
10. Chiţescu, I.: Finitely purely atomic measures and $\mathcal{L}^p$-spaces. An. Univ. Bucureşti Şt. Nat. **24**, 23–29 (1975)
11. Chiţescu, I.: Finitely purely atomic measures: coincidence and rigidity properties. Rend. Circ. Mat. Palermo (2) **50**(3), 455–476 (2001)
12. Dinculeanu, N.: Measure Theory and Real Functions (in Romanian). Didactic and Pedagogical Publishing House, Bucharest (1964)
13. Drewnowski, L.: Topological rings of sets, continuous set functions, integration, I, II, III. Bull. Acad. Polon. Sci. **20**, 269–276, 277–286, 439–445 (1972)
14. Flanders, H.: Differential Forms with Applications to the Physical Science. Dover Publication, New York (1989)
15. Gavriluţ, A.: Non-atomicity and the Darboux property for fuzzy and non-fuzzy Borel/Baire multivalued set functions. Fuzzy Sets and Systems 160 (2009), 1308–1317. Erratum Fuzzy Sets Syst. **161**, 2612–2613 (2010)
16. Gavriluţ, A.: Fuzzy Gould integrability on atoms. Iran. J. Fuzzy Syst. **8**(3), 113–124 (2011)
17. Gavriluţ, A.: Regular Set Multifunctions. Pim Publishing House, Iaşi (2012)
18. Gavriluţ, A., Agop, M.: An Introduction to the Mathematical World of Atomicity through a Physical Approach. ArsLonga Publishing House, Iaşi (2016)
19. Gavriluţ, A., Croitoru, A.: On the Darboux property in the multivalued case. Ann. Univ. Craiova Math. Comput. Sci. Ser. **35**, 130–138 (2008)
20. Gavriluţ, A., Croitoru, A.: Non-atomicity for fuzzy and non-fuzzy multivalued set functions. Fuzzy Sets Syst. **160**, 2106–2116 (2009)
21. Gavriluţ, A., Croitoru, A.: Pseudo-atoms and Darboux property for set multifunctions. Fuzzy Sets Syst. **161**(22), 2897–2908 (2010)
22. Gavriluţ, A., Iosif, A., Croitoru, A.: The Gould integral in Banach lattices. Positivity **19**(1), 65–82 (2015)
23. Gregori, V., Miñana, J.-J., Morillas, S., Sapena, A.: Cauchyness and convergence in fuzzy metric spaces. Rev. Real Acad. Cienc. Exact. Fís. Natur. Ser. A Mat. **111**(1), 25–37 (2017)
24. Grigorovici, A., Băcăiţă, E.S., Păun, V.P., Grecea, C., Butuc, I., Agop, M., Popa, O.: Pairs generating as a consequence of the fractal entropy: theory and applications. Entropy **19**, 128 (2017). https://doi.org/10.3390/e19030128
25. Gudder, S.: Quantum measure and integration theory. J. Math. Phys. **50**, 123509 (2009). https://doi.org/10.1063/1.3267867
26. Gudder, S.: Quantum integrals and anhomomorphic logics, arXiv: quant-ph (0911.1572) (2009)
27. Gudder, S.: Quantum measure theory. Math. Slovaca **60**, 681–700 (2010)
28. Gudder, S.: Quantum measures and the coevent interpretation. Rep. Math. Phys. **67**, 137–156 (2011)
29. Gudder, S.: Quantum measures and integrals, arXiv:1105.3781 (2011)
30. Hartle, J.B.: The quantum mechanics of cosmology. In: Lectures at Winter School on Quantum Cosmology and Baby Universes, Jerusalem, Israel, Dec 27, 1990–Jan 4, 1990 (1990)
31. Hartle, J.B.: Spacetime quantum mechanics and the quantum mechanics of spacetime. In: Zinn-Justin, J., Julia, B. (eds.) Proceedings of the Les Houches Summer School on Gravitation and Quantizations, Les Houches, France, 6 Jul–1, Aug 1992. North-Holland, Amsterdam (1995), arXiv:gr-qc/9304006

32. Iannaccone, P.M., Khokha, M.: Fractal Geometry in Biological Systems: An Analitical Approach. CRC Press, New York (1995)
33. Jaynes, E.T.: The well posed problem. Found. Phys. **3**, 477–493 (1973)
34. Khare, M., Singh, A.K.: Atoms and Dobrakov submeasures in effect algebras. Fuzzy Sets Syst. **159**(9), 1123–1128 (2008)
35. Li, J., Mesiar, R., Pap, E.: Atoms of weakly null-additive monotone measures and integrals. Inf. Sci. 257, 183–192 (2014)
36. Li, J., Mesiar, R., Pap, E., Klement, E.P.: Convergence theorems for monotone measures. Fuzzy Sets Syst. **281**, 103–127 (2015)
37. Mandelbrot, B.B.: The Fractal Geometry of Nature, Updated and augm. edn. W.H. Freeman, New York (1983)
38. Mazilu, N., Agop, M.: Skyrmions. A Great Finishing Touch to Classical Newtonian Philosophy. World Philosophy Series. Nova Science Publishers, New York (2011)
39. Mercheş, I., Agop, M.: Differentiability and Fractality in Dynamics of Physical Systems. World Scientific, Singapore (2015)
40. Mesiar, R., Li, J., Ouyang, Y.: On the equality of integrals. Inf. Sci. **393**, 82–90 (2017)
41. Michel, O.D., Thomas, B.G.: Mathematical Modeling for Complex Fluids and Flows. Springer, New York (2012)
42. Mihăileanu, M.: Differential, Projective and Analytical Geometry (in Romanian). Didactic and Pedagogical Publishing House, Bucharest (1972)
43. Mitchell, M.: Complexity: A Guided Tour. Oxford University Press, Oxford (2009)
44. Nottale, L.: Scale Relativity and Fractal Space-Time. A New Approach to Unifying Relativity and Quantum Mechanics. Imperial College Press, London (2011)
45. Ouyang, Y., Li, J., Mesiar, R.: Relationship between the concave integrals and the pan-integrals on finite spaces. J. Math. Anal. Appl. **424**, 975–987 (2015)
46. Pap, E.: The range of null-additive fuzzy and non-fuzzy measures. Fuzzy Sets Syst. **65**(1), 105–115 (1994)
47. Pap, E.: Null-additive Set Functions. Mathematics and its Applications, vol. 337. Springer, Berlin (1995)
48. Pap, E.: Some Elements of the Classical Measure Theory. Chapter 2 in Handbook of Measure Theory, pp. 27–82. Elsevier, Amsterdam (2002)
49. Pap, E., Gavriluţ, A., Agop, M.: Atomicity via regularity for non-additive set multi functions. Soft Comput. (Foundations) **20**, 4761–4766 (2016). https://doi.org/10.1007/s00500-015-2021-x
50. Phillips, A.C.: Introduction to Quantum Mechanics. Wiley, New York (2003)
51. Rao, K.P.S.B., Rao, M.B.: Theory of Charges. Academic, New York (1983)
52. Salgado, R.: Some identities for the q-measure and its generalizations. Mod. Phys. Lett. A **17**, 711–728 (2002)
53. Schweizer, B., Sklar, A.: Probabilistic Metric Spaces. Elsevier Science, New York (1983). Republished in 2005 by Dover Publications, Inc., with a new preface, errata, notes, and supplementary references
54. Sorkin, R.D.: Quantum mechanics as quantum measure theory. Mod. Phys. Lett. A **9**, 3119–3128 (1994), arXiv:gr-qc/9401003
55. Sorkin, R.D.: Quantum measure theory and its interpretation. In: Feng, D.H., Hu, B.-L. (eds.) Quantum Classical Correspondence: Proceedings of the 4th Drexel Symposium on Quantum Non-integrability, pp. 229–251. International Press, Cambridge (1997)
56. Sorkin, R.: Quantum dynamics without the wave function. J. Phys. A Math. Theor. **40**, 3207–3231 (2007)
57. Surya, S., Waldlden, P.: Quantum covers in q-measure theory, arXiv: quant-ph 0809.1951 (2008)
58. Susskind, L.: Copenhagen vs Everett teleportation, and ER ≡ EPR, arXiv: 1604.02589v2 (23 April 2016)

59. Susskind, L.: ER ≡ EPR,GHZ, and the consistency of quantum measurements. Fortsch. Phys. **64**, 72 (2016)
60. Suzuki, H.: Atoms of fuzzy measures and fuzzy integrals. Fuzzy Sets Syst. **41**, 329–342 (1991)
61. Wu, C., Bo, S.: Pseudo-atoms of fuzzy and non-fuzzy measures. Fuzzy Sets Syst. **158**, 1258–1272 (2007)

Chapter 11
On a multifractal theory of motion in a non-differentiable space: Toward a possible multifractal theory of measure

11.1 Introduction

Since the non-differentiability becomes a fundamental property of the motions space, a correspondence between the interaction processes and multifractality of the motion trajectories can be established. Then, for all scale resolutions, the geodesics equations (either in the form of the Schrödinger equation of fractal type or in the form of the hydroynamic equations of fractal type) and some applications are obtained. All these results specify the fact at a multifractal measure theory is absolutely necessary to be built in order to be able to describe nature's dynamics.

11.2 Consequences of non-differentiability on a space manifold

Let us assume that the motions of the physical systems take place on continuous but non-differentiable curves (fractal curves), so that the following consequences are resulting [3, 5]:

(i) Any continuous but non-differentiable curve of the physical systems is explicitly scale resolution δt dependent, i.e., its length tends to infinity when δt tends to zero;

We mention that, mathematically speaking, a curve is non-differentiable if it satisfies the Lebesgue theorem [4], i.e., its length becomes infinite when the scale resolution goes to zero. Consequently, in this limit, a curve is as zig-zagged as one can imagine. Thus, it exhibits the property of self-similarity in every one of its points, which can be translated into a property of holography (every part reflects the whole) [4, 5]. This concept of holography can lead to

A. Gavriluţ et al., *Atomicity through Fractal Measure Theory*,
https://doi.org/10.1007/978-3-030-29593-6_11

new models for the evolution of cancer, or new models of neural interactions etc.;

(ii) The physics of phenomena is related to the behaviour of a set of functions during the zoom operation of the scale resolution δt. Then, through the substitution principle, δt will be identified with dt, i.e., $\delta t \equiv dt$ and, consequently, it will be considered as an independent variable. We reserve the notation dt for the usual time as in the Hamiltonian physical system dynamics;

(iii) The physical system dynamics is described through fractal variables, i.e., functions depending on both the space coordinates and the scale resolution since the differential time reflection invariance of any dynamical variable is broken. Then, in any point of a physical system fractal curve, two derivatives of the variable field $Q(t, dt)$ can be defined:

$$\frac{d_+ Q(t,dt)}{dt} = \lim_{\Delta t \to 0_+} \frac{Q(t+\Delta t, \Delta t) - Q(t, \Delta t)}{\Delta t}$$

$$\frac{d_- Q(t,dt)}{dt} = \lim_{\Delta t \to 0_-} \frac{Q(t, \Delta t) - Q(t-\Delta t, \Delta t)}{\Delta t}. \tag{11.1}$$

The "+" sign corresponds to the physical system forward processes, while the "−" sign corresponds to the backwards ones;

(iv) The differential of the spatial coordinate field $dX^i(t, dt)$, by means of which we can describe the physical system dynamics, is expressed as the sum of the two differentials, one of them being scale resolution independent (differential part $d_\pm x^i(t)$, and the other one being scale resolution dependent (fractal part $d_\pm \xi^i(t)$), i.e.,

$$d_\pm X^i(t, dt) = d_\pm x^i(t) + d_\pm \xi^i(t, dt); \tag{11.2}$$

(v) The non-differentiable part of the spatial coordinate field, by means of which we can describe the physical system dynamics, satisfies the fractal equation:

$$d_\pm \xi^i(t, dt) = \lambda^i_\pm (dt)^{1/D_F} \tag{11.3}$$

where $\lambda^i_\pm$ are constant coefficients through which the fractalization type describing the physical system dynamics is specified and D_F defines the fractal dimension of the physical system non-differentiable curve.

In our opinion, physical systems processes imply dynamics on geodesics with various fractal dimensions. The variety of these fractal dimensions of the physical systems geodesics comes as a result of its structure. Precisely, for $D_F = 2$, quantum type processes are generated in a physical system. For $D_F < 2$ correlative type processes are induced, while for $D_F > 2$ non-correlative type ones can be found—for details see [5].

Because all the processes described here can take place simultaneously in the dynamics of a physical system, it is thus necessary to consider the multifractal behaviour of physical structures;

(vi) The differential time reflection invariance of any dynamical variable of the physical system is recovered by combining the derivatives d_+/dt and d_-/dt in the non-differentiable operator:

$$\frac{\hat{d}}{dt} = \frac{1}{2}\left(\frac{d_+ + d_-}{dt}\right) - \frac{i}{2}\left(\frac{d_+ - d_-}{dt}\right). \tag{11.4}$$

This is a natural result of the complex prolongation procedure applied to physical system dynamics [1]. Applying now the non-differentiable operator to the spatial coordinate field, by means of which we can describe the physical system dynamics, yields the complex velocity field:

$$\hat{V}^i = \frac{\hat{d}X^i}{dt} = V_D^i - V_F^i \tag{11.5}$$

with

$$V_D^i = \frac{1}{2}\frac{d_+X^i + d_-X^i}{dt},\ V_F^i = \frac{1}{2}\frac{d_+X^i - d_-X^i}{dt}. \tag{11.6}$$

The real part V_D^i of the complex velocity field is differentiable and scale resolution independent (differentiable velocity field), while the imaginary one V_F^i is non-differentiable and scale resolution dependent (fractal velocity field);

(vii) In the absence of any external constraint, an infinite number of fractal curves (geodesics) can be found relating any pair of points, and this is true on all scales of the physical system dynamics. Then, in the fractal space of the physical system, all its entities are substituted with the geodesics themselves so that any external constraint can be interpreted as a selection of geodesics. The infinity of geodesics in the bundle, their non-differentiability and the two values of the derivative imply a generalized statistical fluid-like description (in what follows we shall call it a fractal fluid). Then, the average values of the fractal fluid variables must be considered in the previously mentioned sense, so the average of $d_\pm X^i$ is:

$$< d_\pm X^i >\equiv d_\pm x^i \tag{11.7}$$

with

$$< d_\pm \xi^i >= 0. \tag{11.8}$$

The previous relation (11.8) implies that the average of the fractal fluctuations is null.

(viii) The fractal fluid dynamics can be described through a scale covariant derivative, the explicit form of which is obtained as follows. Let us consider that the non-differentiable curves are immersed in a 3-dimensional space and that X^i are the spatial coordinate field of a point on the non-differentiable curve. We also consider a variable field $Q(X^i, t)$ and the following Taylor expansion up to the second order:

$$d_\pm Q(X^i, t) = \partial_t Q dt + \partial_i Q d_\pm X^i + \frac{1}{2}\partial_l \partial_k Q d_\pm X^l d_\pm X^k + \cdots . \tag{11.9}$$

These relations are valid in any point and more for the points X^i on the non-differentiable curve which we have selected in (11.9). From here, the main forward and backward values for fractal fluid variables from (11.9) become:

$$< d_\pm Q >=< \partial_t Q dt > + < \partial_i Q d_\pm X^i > + \frac{1}{2} < \partial_l \partial_k Q d_\pm X^l d_\pm X^k > + \cdots . \tag{11.10}$$

We suppose that the average values of the all variable field Q and its derivatives coincide with themselves and the differentials $d_\pm X^i$ and dt are independent. Therefore, the average of their products coincides with the product of averages. Consequently, (11.10) becomes:

$$d_\pm Q = \partial_t Q dt + \partial_i Q < d_\pm X^i > + \frac{1}{2}\partial_l \partial_k Q < d_\pm X^l d_\pm X^k > + \cdots . \tag{11.11}$$

Even the average value of $d_\pm \xi^i$ is null, for the higher order of $d_\pm \xi^i$ the situation can still be different. Let us focus on the averages $< d_\pm \xi^l d_\pm \xi^k >$. Using (11.3) we can write:

$$< d_\pm \xi^l d_\pm \xi^k > \pm \lambda_\pm^l \lambda_\pm^k (dt)^{(2/D_F)-1} dt, \tag{11.12}$$

where we accepted that the sign $+$ corresponds to $dt > 0$ and the sign $-$ corresponds to $dt < 0$.

Then, (11.11) takes the form:

$$d_\pm Q = \partial_t Q dt + \partial_i Q < d_\pm X^i > + \frac{1}{2}\partial_l \partial_k Q d_\pm x^l d_\pm x^k$$
$$\pm \frac{1}{2}\partial_l \partial_k Q[\lambda_\pm^l \lambda_\pm^k (dt)^{(2/D_F)-1} dt]. \tag{11.13}$$

If we divide by dt and neglect the terms that contain differential factors (for details, see the method from [3, 5]) we obtain:

$$\frac{d_{\pm}Q}{dt} = \partial_t Q + v_{\pm}^i \partial_i Q \pm \frac{1}{2}\lambda_{\pm}^l \lambda_{\pm}^k (dt)^{(2/D_F)-1} \partial_l \partial_k Q. \tag{11.14}$$

These relations also allow us to define the operators

$$\frac{d_{\pm}}{dt} = \partial_t + v_{\pm}^i \partial_i \pm \frac{1}{2}\lambda_{\pm}^l \lambda_{\pm}^k (dt)^{(2/D_F)-1} \partial_l \partial_k. \tag{11.15}$$

where

$$v_+^i = \frac{d_+ x^i}{dt},\; v_-^i = \frac{d_- x^i}{dt}.$$

Under these circumstances, taking into account (11.4), (11.5) and (11.15), let us calculate $\hat{d}/dt$. It results:

$$\frac{\hat{d}Q}{dt} = \partial_t Q + \hat{V}^i \partial_i Q + \frac{1}{4}(dt)^{(2/D_F)-1} D^{lk} \partial_l \partial_k Q. \tag{11.16}$$

where

$$\begin{aligned} D^{lk} &= d^{lk} - i\bar{d}^{lk} \\ d^{lk} &= \lambda_+^l \lambda_+^k - \lambda_-^l \lambda_-^k, \bar{d}^{lk} = \lambda_+^l \lambda_+^k + \lambda_-^l \lambda_-^k. \end{aligned} \tag{11.17}$$

The relation (11.16) also allows us to define the scale covariant derivative in the fractal fluid dynamics

$$\frac{\hat{d}}{dt} = \partial_t + \hat{V}^i \partial_i + \frac{1}{4}(dt)^{(2/D_F)-1} D^{lk} \partial_l \partial_k. \tag{11.18}$$

11.2.1 Fractal fluid geodesics

Let us now consider the principle of scale covariance (the physics laws, which are specific to the physical fractal fluid dynamics, are invariant with respect to scale transformations) and postulate that the passage from the classical (differentiable) physics to the fractal (non-differentiable) physics can be implemented by replacing the standard time derivative d/dt with the non-differentiable operator $\hat{d}/dt$. Thus, this operator plays the role of the scale covariant derivative, namely it is used to write the fundamental equations of the fractal fluid dynamics in the same form as in the classic (differentiable) case. Under these conditions, applying the operator (11.18) to the complex velocity field (11.5), in the absence of any external constraint, the geodesics take the following form:

$$\frac{\hat{d}\hat{V}^i}{dt} = \partial_t \hat{V}^i + \hat{V}^l \partial_l \hat{V}^i + \frac{1}{4}(dt)^{(2/D_F)-1} D^{lk} \partial_l \partial_k \hat{V}^i = 0. \tag{11.19}$$

This means that the local acceleration $\partial_t \hat{V}^i$, the local convection $\hat{V}^l \partial_i \hat{V}^i$ and the local dissipation $D^{lk} \partial_l \partial_k \hat{V}^i$, make their balance in any point of the non-differentiable curve. Moreover, the presence of the complex coefficient of viscosity-type in $\frac{1}{4}(dt)^{(2/D_F)-1} D^{lk}$ in the fractal fluid dynamics specifies that it is a rheological medium. So, it has memory, as a datum, by its own structure.

If the fractalization is achieved by Markov type stochastic processes, which involve Lévy type movements [2] of the fractal fluid entities, then:

$$\lambda_+^i \lambda_+^l = \lambda_-^i \lambda_-^l = 2\lambda \delta^{il}, \tag{11.20}$$

where δ^{il} is the Kronecker's pseudo-tensor.

Under these conditions, the geodesic equation of the fractal fluid takes the simple form

$$\frac{\hat{d}\hat{V}^i}{dt} = \partial_t \hat{V}^i + \hat{V}^l \partial_l \hat{V}^i - i\lambda(dt)^{(2/D_F)-1} \partial^l \partial_l \hat{V}^i = 0 \tag{11.21}$$

or more, by separating the motions on differential and fractal scale resolutions,

$$\begin{aligned} \frac{\hat{d}V_D^i}{dt} &= \partial_t V_D^i + V_D^l \partial_l V_D^i - [V_F^l + \lambda(dt)^{(2/D_F)-1} \partial^l] \partial_l V_F^i = 0 \\ \frac{\hat{d}V_F^i}{dt} &= \partial_t V_F^i + V_D^l \partial_l V_F^i + [V_F^l + \lambda(dt)^{(2/D_F)-1} \partial^l] \partial_l V_D^i = 0. \end{aligned} \tag{11.22}$$

11.3 Fractality and its implications

The separation of the physical system dynamics specifies at non-differentiable resolution scales the fractal force:

$$F_F^i = (V_F^l + \lambda(dt)^{(2/D_F)-1} \partial^l) \partial_l V_F^i. \tag{11.23}$$

Its cancellation

$$(V_F^l + \lambda(dt)^{(2/D_F)-1} \partial^l) \partial_l V_F^i = 0 \tag{11.24}$$

in the condition

$$\partial_l V_F^l = 0 \tag{11.25}$$

induces a particular field of velocities whose explicit form will be given in the following.

Finding the solutions for these equations can be relatively difficult, due to the fact that this equation system is a non-linear one. However, there is an analytical solution of this system, in the particular case of a "stationary flow" in a plane symmetry (x, y). In these circumstances, equations (11.24) and (11.25) take the form:

$$V_x \frac{\partial V_x}{\partial x} + V_y \frac{\partial V_x}{\partial x} = \lambda (dt)^{(2/D_F)-1} \frac{\partial^2 V_x}{\partial y^2} \tag{11.26}$$

$$\frac{\partial V_x}{\partial x} + \frac{\partial V_y}{\partial y} = 0, \tag{11.27}$$

where $V_{Fx} = V_x(x, y)$ is the velocity along axis Ox, $V_{Fy} = V_y(x, y)$ is the velocity along axis Oy. The boundary condition of the flow are:

$$\lim_{y \to 0} V_y(x, y) = 0, \ \lim_{y \to 0} \frac{\partial V_x}{\partial y} = 0, \ \lim_{y \to \infty} V_x(x, y) = 0 \tag{11.28}$$

and the flux momentum per length unit is constant:

$$\Theta = \rho \int_{-\infty}^{+\infty} V_x^2 dy = \text{const.} \tag{11.29}$$

Using the method from [7] for solving the equations (11.26) and (11.27), with the limit conditions (11.28) and (11.29), the following solutions result:

$$V_x = \frac{[1,5(\frac{\Theta}{6\rho})^{2/3}]}{[\lambda(dt)^{(2/D_F)-1}x]^{1/3}} \cdot \sec h^2 \frac{[(0,5y)(\frac{\Theta}{6\rho})^{1/3}]}{[\lambda(dt)^{(2/D_F)-1}x]^{2/3}} \tag{11.30}$$

$$V_y = \frac{[4,5(\frac{\Theta}{6\rho})^{2/3}]}{[3\lambda(dt)^{(2/D_F)-1}x]^{1/3}} \cdot \left[\frac{y(\frac{\Theta}{6\rho})^{1/3}}{[\lambda(dt)^{(2/D_F)-1}x]^{2/3}} \cdot \right.$$

$$\left. \cdot \sec h^2 \frac{[(0,5y)(\frac{\Theta}{6\rho})^{1/3}]}{[\lambda(dt)^{(2/D_F)-1}x]^{2/3}} - \tanh \frac{[(0,5y)(\frac{\Theta}{6\rho})^{1/3}]}{[\lambda(dt)^{(2/D_F)-1}x]^{2/3}} \right] \tag{11.31}$$

Relations (11.30) and (11.31) suggest that at all scale resolutions, the fractal fluid velocity field is highly non-linear by means of soliton and soliton-kink type solutions.

For $y = 0$, we obtain in relation (11.30) the flow critical velocity in the form:

$$V_x(x, y = 0) = V_c = \frac{[1,5(\frac{\Theta}{6\rho})^{2/3}]}{[\lambda(dt)^{(2/D_F)-1}x]^{1/3}} \tag{11.32}$$

while relation (11.29), taking into account (11.32) becomes:

$$\Theta = \rho \int_{-\infty}^{+\infty} V_x^2(x, y)dy = \int_{-d_c}^{+d_c} V_c^2(x, 0)dy \tag{11.33}$$

so that the critical cross section of the strains lines tube is given by:

$$d_c(x, y = 0) = \frac{\Theta}{2\rho V_c^2} = 2,42[\lambda(dt)^{(2/D_F)-1}x]^{2/3}(\frac{\rho}{\Theta})^{1/3}. \tag{11.34}$$

11.4 Fractal geodesics in the Schrödinger type representation

For irrotational motions of the fractal fluid, the complex velocity field $\hat{V}^l$ takes the form:

$$\hat{V}^l = -2i\lambda(dt)^{(2/D_F)-1}\partial^l \ln\Psi. \tag{11.35}$$

Then the geodesics equation (11.21) becomes:

$$\begin{aligned}\frac{\hat{d}\hat{V}^l}{dt} &= -2i\lambda(dt)^{(2/D_F)-1}\partial_t\partial^l \ln\Psi \\ &+ [-2i\lambda(dt)^{(2/D_F)-1}\partial_p \ln\Psi - i\lambda(dt)^{(2/D_F)-1}\partial_p] \\ &\times \partial^p\partial^l[-2i\lambda(dt)^{(2/D_F)-1}\ln\Psi] = 0.\end{aligned} \tag{11.36}$$

Since

$$\begin{aligned}\partial^l(\partial_p \ln\Psi\partial^p \ln\Psi) &= 2\partial_p \ln\Psi\partial^p\partial^l \ln\Psi \\ \partial^l\partial_p\partial^p \ln\Psi &= \partial_p\partial^p\partial^l \ln\Psi \\ \partial^l(\partial_p \ln\Psi\partial^p \ln\Psi) + \partial_p\partial^p \ln\Psi) &= \partial^l\left(\frac{\partial_p\partial^p\Psi}{\Psi}\right)\end{aligned} \tag{11.37}$$

equation (11.36) becomes:

$$i\lambda(dt)^{(2/D_F)-1}\partial_t\partial^l \ln\Psi + \lambda^2(dt)^{(4/D_F)-2}\partial^l\left(\frac{\partial_p\partial^p\Psi}{\Psi}\right) = 0. \tag{11.38}$$

By integrating the above relation we obtain:

$$\lambda^2(dt)^{(4/D_F)-2}\partial_p\partial^p\Psi + i\lambda(dt)^{(2/D_F)-1}\partial_t\Psi + Q^2(t)\Psi = 0 \tag{11.39}$$

with $Q^2(t)$ an arbitrary function of t.

In consequence, the non-differentiable geodesics in the terms of Ψ are defined up to an arbitrary function $Q^2(t)$, which is dependent on t. For $Q(t) \equiv 0$, the relation (11.39) reduces to the fractal type Schrödinger equation:

$$\lambda^2(dt)^{(4/D_F)-2}\partial_p\partial^p\Psi + i\lambda(dt)^{(2/D_F)-1}\partial_t\Psi = 0. \tag{11.40}$$

The standard Schrödinger equation:

$$\frac{\hbar^2}{2m_0}\partial_p\partial^p\Psi + i\hbar\partial_t\Psi = 0$$

can be obtained from (11.40) for non-relativistic motions on Peano curves [5], $D_F = 2$, at Compton scale $\lambda = \hbar/2m_0$.

In the case of non-differentiable dynamics with constraints, for instance, under the action of a scalar potential U, following the same procedure as before, one obtains the fractal type equation:

$$\lambda^2(dt)^{(4/D_F)-2}\partial^l\partial_l\Psi + i\lambda(dt)^{(2/D_F)-1}\partial_t\Psi - \frac{U}{2}\Psi = 0. \tag{11.41}$$

For non-relativistic motions on Peano curves, $D_F = 2$, at Compton scale $\lambda = \hbar/2m_0$ (for details see [5]) the equation (11.40) takes the standard form:

$$\frac{\hbar^2}{2m_0}\partial_l\partial^l\Psi + i\hbar\partial_t\Psi - U\Psi = 0.$$

If the scalar potential U is time independent, the equation (11.40) admits the stationary solution:

$$\Psi(\mathbf{r}, t) = \Psi(\mathbf{r}) \exp\left[-\frac{i}{2m_0\lambda(dt)^{(2/D_F)-1}}Et\right], \tag{11.42}$$

where E is the fractal energy of the physical system of fractal type. Then $\Psi(\mathbf{r})$ is the solution of non-temporal Schrödinger equation of fractal type:

$$\partial_l\partial^l\Psi + \frac{1}{2m_0\lambda^2(dt)^{(4/D_F)-2}}(E - U)\Psi = 0. \tag{11.43}$$

11.5 The one dimensional potential barrier: Fractal tunnel type effect

Let us consider a one dimensional potential barrier of rectangular form for which the potential has the form:

$$U(X) = \begin{cases} 0, & -\infty < X < 0 \\ U_0, & 0 < X < a \\ 0, & a < X < +\infty \end{cases}.$$

As one can easily observe from the potential form, the real straight line $\{X; X \in \mathbb{R}\}$ is structured in three regions, denoted by 1, 2, 3 and called, of incidence, of barrier and of emergence, respectively. The energy E of the physical system was deliberately chosen smaller than the energy U_0, in the barrier region, just in order to "simulate" tunnel effect of the fractal type.

Denoting by $\theta_1, \theta_2, \theta_3$ the functions ($R \to C$, defined on R wth values in C) corresponding to the fractal states of the physical system into the above-mentioned three regions, we have the following equations of the stationary Schrödinger fractal type equation:

$$\begin{aligned} &\frac{d^2\theta_1}{dX^2} + k^2\theta_1 = 0, -\infty < X < 0 \\ &\frac{d^2\theta_2}{dX^2} - q^2\theta_2 = 0, 0 < X < a \\ &\frac{d^2\theta_3}{dX^2} + k^2\theta_3 = 0, a < X < +\infty. \end{aligned} \tag{11.44}$$

In equations (11.44), for convenience, we have made the notations:

$$k^2 = \frac{E}{2m_0\lambda^2(dt)^{(4/D_F)-2}}, q^2 = \frac{U_0 - E}{2m_0\lambda^2(dt)^{(4/D_F)-2}}. \tag{11.45}$$

The simple form of the above three differential equations leads to their quick integration and the following solutions are obtained as:

$$\begin{aligned} &\theta_1(X) = A_1e^{ikX} + B_1e^{-ikX}, -\infty < X < 0 \\ &\theta_2(X) = A_2e^{qX} + B_2e^{-qX}, 0 < X < a \\ &\theta_3(X) = A_3e^{ikX}, a < X < +\infty, \end{aligned} \tag{11.46}$$

where A_1, B_1, A_2, B_2, A_3 are constants from C. The fractal modes $\{e^{ikX}; k \in \mathbb{R}_+\}$ are associated with the direct fractal wave which is incident (from $-\infty$) in the region 1 and emergent (to $+\infty$) in the region 3. The fractal modes $\{e^{ikX}; k \in \mathbb{R}_+\}$ are associated with the reflected fractal wave which exists only in the region 1, passing from $X = 0$ to $X = -\infty$, since in the region 3 the potential is uniform null. The justification of this interpretation is based on expression of the fractal states density current:

$$J_X = i\lambda(dt)^{(2/D_F)-1}\left(\theta\frac{d\overline{\theta}}{dX} - \overline{\theta}\frac{d\theta}{dX}\right) \quad (11.47)$$

which for the fractal direct waves $A_{(1,3)}e^{ikX}$ becomes:

$$J_{(1,3)} = 2\lambda(dt)^{(2/D_F)-1}|A_{(1,3)}|^2. \quad (11.48)$$

It is positive defined and represents the density of the incident fractal states density current in region 1:

$$J_1 = 2\lambda(dt)^{(2/D_F)-1}k|A_1|^2 \quad (11.49)$$

and emergent in region 3:

$$J_e = 2\lambda(dt)^{(2/D_F)-1}k|A_3|^2. \quad (11.50)$$

For the fractal reflected wave, of amplitude B_1, described by fractal mode e^{-ikX}, we have:

$$J_r = -2\lambda(dt)^{(2/D_F)-1}k|B_1|^2 \quad (11.51)$$

i.e., is indeed resulting a fractal states density current, oriented to $-\infty$, associated with the physical system reflected by the wall $X = 0$ of the potential barrier. This leads to the possibility of an univocal characterization of the tunnel effect of fractal type, through the fractal transparency:

$$T = \frac{J_e}{J_i} = |\frac{A_3}{A_1}|^2 \quad (11.52)$$

and the fractal reflectance

$$R = \frac{J_r}{J_i} = |\frac{B_1}{A_1}|^2. \quad (11.53)$$

As these values are independent from the constants A_2 and B_2, their direct estimation presents no direct interest. Therefore, imposing the coupling conditions (in $X = 0$ and $X = a$), for θ_i functions with $i = 1, 2, 3$ and for their derivatives, which means:

$$\theta_1(0) = \theta_2(0)$$
$$\frac{d\theta_1}{dX}(0) = \frac{d\theta_2}{dX}(0)$$
$$\theta_2(a) = \theta_3(a)$$

$$\frac{d\theta_2}{dX}(a) = \frac{d\theta_3}{dX}(a) \tag{11.54}$$

the algebraic system is obtained as:

$$\begin{aligned} A_1 + B_1 &= A_2 + B_2 \\ ik(A_1 - B_1) &= q(A_2 - B_2) \\ e^{qa} A_2 + e^{-qa} B_2 &= e^{iqa} A_3 \\ q(e^{qa} A_2 + e^{-qa} B_2 &= ike^{iqa} A_3. \end{aligned} \tag{11.55}$$

In this algebraic system we will seek the elimination of the unknown quantities A_2 and B_2 to obtain the expressions:

$$\tau = \frac{A_3}{A_1}, r = \frac{B_1}{A_1} \tag{11.56}$$

called the complex fractal transmission factor, τ, respectively, complex fractal reflection factor, r, because the following relationships are satisfied:

$$\begin{aligned} T &= \overline{\tau}\tau = |\tau|^2 \\ R &= \overline{r}r = |r|^2. \end{aligned} \tag{11.57}$$

By dividing the last two equations from (11.55), member by member, it results:

$$\frac{B_2}{A_2} = \frac{q - ik}{q + ik} e^{2qa} \tag{11.58}$$

which substituted in the ratio of the first equations from (11.55), i.e.:

$$\frac{1-r}{1+r} = -i\frac{q}{k}\frac{1 - B_2/A_2}{1 + A_2/A_2} \tag{11.59}$$

leads, through solving in relation with r and by the appropriate grouping of terms, to the expression of the complex fractal reflection factor:

$$r = -\frac{k^2 + q^2}{(q^2 - k^2) - 2iqk \cdot cth(qa)} \tag{11.60}$$

and to the expression form of the complex fractal reflectance factor:

$$R = \frac{(k^2 + q^2)^2}{(q^2 - k^2) + 4q^2k^2 \cdot cth^2(qa)}. \tag{11.61}$$

It is noted that R has values between 0 (for $qa \to 0$) and 1 (for $a \to \infty$), specifically the fractal reflection of the physical system on the potential barrier.

Now, based on the conservation law of the fractal states density current:

$$J_i + J_r = J_e \tag{11.62}$$

explicitly, we have:

$$2\lambda(dt)^{(2/D_F)-1}(|A_1|^2 - |B_1|^2) = 2\lambda(dt)^{(2/D_F)-1}|A_3|^2. \tag{11.63}$$

Considering the way we defined the fractal transparency, and the fractal reflectance of barrier, respectively, from the above relation it can be found:

$$T + R \equiv 1. \tag{11.64}$$

From the first equations of the algebraic system (11.55), removing the unknown B_2, it can be obtained as:

$$\frac{A_2}{A_1} = (2q)^{-1}[(q+ik) + r(q-ik)] \tag{11.65}$$

while from the third equation of the same algebraic system, it can be found:

$$\tau = e^{-ika}\left[\frac{A_2}{A_1}e^{qa} + \frac{B_2}{A_2}e^{-qa}\right]. \tag{11.66}$$

In these conditions the complex fractal transmission factor can be written as:

$$\tau = -\frac{2iqke^{-iqa}}{(q^2-k^2)sh(qa) - 2iqk \cdot ch(qa)}. \tag{11.67}$$

From here, the fractal transparency becomes:

$$T = \frac{4q^2k^2}{4q^2k^2 + (q^2+k^2)sh^2(qa)} \tag{11.68}$$

having, as well, values between 1 (for $qa \to 0$) and 0 (for $a \to \infty$).

Finally, using the notations (11.45), we obtain the final form of the relations (11.61) and (11.68), i.e.,

$$R = \frac{U_0^2 sh^2\left\{\left[\frac{U_0 - E}{2m_0\lambda^2(dt)^{(4/D_F)-2}}\right]^{1/2} a\right\}}{U_0^2 sh^2\left\{\left[\frac{U_0 - E}{2m_0\lambda^2(dt)^{(4/D_F)-2}}\right]^{1/2} a\right\} + 4E(U_0 - E)} \tag{11.69}$$

$$T = \frac{4E(U_0 - E)}{U_0^2 sh^2 \left\{ \left[\frac{U_0 - E}{2m_0 \lambda^2 (dt)^{(4/D_F)-2}} \right]^{1/2} a \right\} + 4E(U_0 - E)}. \tag{11.70}$$

These results prove the fact that there do not exist particles that wander indefinitely, by multiple reflections, inside the tunnel area, so that the particles could not be found in the emerging region of probability T, or in the incidence region, as reflected particles, of probability R.

11.6 Fractal motions in central field

Let us consider the case of a field of the form

$$U(r) = -\frac{C}{r}, C = \text{const.}$$

Since this field has spherical symmetry, the fractal Schrödinger type stationary equation (11.43) becomes:

$$\frac{\partial}{\partial r}\left(r^2 \frac{\partial \Psi}{\partial r}\right) + \frac{1}{2m_0 \lambda^2 (dt)^{(4/D_F)-2}}\left(E + \frac{C}{r}\right)\Psi$$
$$= -\left[\frac{1}{\sin\theta}\frac{\partial}{\partial\theta}\left(\sin\theta \frac{\partial\Psi}{\partial\theta}\right) + \frac{1}{\sin^2\theta}\frac{\partial^2\Psi}{\partial\varphi^2}\right]. \tag{11.71}$$

Following a procedure similar to the one from [6], one obtains the eigen solution:

$$\Psi_{nlm}(r, \theta, \varphi) = R_{nl}(r) Y_{ml}(\theta, \varphi) \tag{11.72}$$

with

$$R_{nl}(r) = \left(\frac{2}{nr_0}\right)^{3/2} \left\{\frac{(n-l-1)!}{2n[(n+l)!]^3}\right\}^{1/2} \left(\frac{2r}{nr_0}\right)^l \exp\left(-\frac{r}{nr_0}\right) L_{n+l}^{2l+1}\left(\frac{2r}{nr_0}\right)$$

$$Y_{ml}(\theta, \varphi) = (-1)^k \left[\frac{(l-m)!(2l+1)}{(l+|m|)!4\pi}\right]^{1/2} P_l^m(\cos\theta)\exp(im\varphi)$$

$$r_0 = \frac{4m_0}{C}\lambda^2 (dt)^{(4/D_F)-2} \tag{11.73}$$

$$n = 1, 2, 3, \ldots, l = 0, 1, 2, \ldots, n-1, m = 0, \pm 1, \pm 2, \ldots, \pm l$$

$$k = \begin{cases} m, & \text{for } m \geq 0 \\ 0, & \text{for } m < 0, \end{cases}$$

where L_{n+l}^{2l+1} are the generalized Laguerre polynomials, P_l^m [6] are the generalized Legendre polynomials, n is the principal fractal number, l is the orbital fractal number, m is the magnetic fractal number, respectively, the eigenvalue:

$$E_n = -2m_0\lambda^2(dt)^{(4/D_F)-2}\frac{1}{n^2r_0^2}. \tag{11.74}$$

11.7 Complex system geodesics in the fractal hydrodynamic representation

If $\Psi = \sqrt{\rho}\exp(iS)$ with $\sqrt{\rho}$ the amplitude and S the phase of Ψ, the complex velocity field (1) takes the explicit form

$$\begin{aligned}\hat{V}^i &= -2i\lambda(dt)^{(2/D_F)-1}\partial^i \ln\Psi,\\ V_D^i &= 2\lambda(dt)^{(2/D_F)-1}\partial^i S,\\ V_F^i &= \lambda(dt)^{(2/D_F)-1}\partial^i \ln\rho.\end{aligned} \tag{11.75}$$

Substituting (11.75) in (11.21) and separating the real and imaginary parts, up to an arbitrary phase factor which may be set to zero by a suitable choice of the phase of Ψ, we obtain:

$$\partial_t V_D^i + (V_D^l\partial_l)V_D^i = -\partial^i(Q), \tag{11.76}$$

$$\partial_t\rho + \partial^i(\rho V_D^i) = 0 \tag{11.77}$$

with Q the specific non-differentiable (fractal) potential

$$Q = -2\lambda^2(dt)^{(4/D_F)-2}\frac{\partial^l\partial_l\sqrt{\rho}}{\sqrt{\rho}} = -\frac{V_F^l V_{Fl}}{2} - \lambda(dt)^{(2/D_F)-1}\partial_i V_F^i. \tag{11.78}$$

We note that the presence of an external scalar potential U modifies the equation (11.76) in the form

$$\partial_t V_D^i + (V_D^l\partial_l)V_D^i = -\partial^i(Q+U). \tag{11.79}$$

Equation (11.76) represents the specific momentum conservation law, while equation (11.77) represents the state density conservation law. Equations (11.76) and (11.77) with (11.78) define the fractal hydrodynamic model and imply the following:

(i) Any entities of the physical system is in a permanent interaction with a fractal medium through the specific non-differentiable potential (11.78);

(ii) The complex system can be identified with a fractal fluid, the dynamics of which is described by the fractal hydrodynamic model;

(iii) The fractal velocity field V_F^i does not represent actual motion, but contributes to the transfer of the specific momentum and to the energy focus. This may be seen clearly from the absence of V_F^i from the states density conservation law and from its role in the variational principle [8];

(iv) Any interpretation of the specific fractal potential should take cognizance of the "self" nature of the specific momentum transfer. While the fractal energy is stored in the form of the mass motion and fractal potential energy, some is available elsewhere and only the total is conserved. It is the conservation of the fractal energy and the fractal momentum that ensure fractal reversibility and the existence of fractal eigenstates, but denies a Lévy motion fractal force of interaction with an external medium [2];

(v) Two types of fractal stationary states are to be distinguished:

(a) *Fractal dynamic states*. For $\partial_t = 0$ and $V_D^l \neq 0$, the equations (11.79) give

$$\frac{1}{2}V_D^l V_{Dl} + U + Q = E,$$
$$\rho V_D^i = \varepsilon^{ilk} \partial_l f_k. \tag{11.80}$$

Therefore, the sum of the specific kinetic energy $V_D^l V_{Dl}/2$, external specific scalar potential, U, and specific fractal potential, Q, is invariant, i.e., equal to the integration constant $E \neq E(x^l)$ (see the first equation (11.80)). $E \equiv < E >$ represents the total specific fractal energy of the fractal dynamic physical system. The fractal states density current, ρV_D^l, has no sources (see the second equation (11.80)), i.e., its fractal streamlines are closed.

(b) *Fractal static states*. For $\partial_t = 0$ and $V_D^l = 0$, the equations (11.79) give

$$U + Q = E. \tag{11.81}$$

The sum of the external specific scalar potential, U, and the specific fractal potential, Q, is invariant, i.e., it is equal to the integration constant $E \neq E(x^l)$ (see equation (11.81)). $E \equiv < E >$ represents the total specific fractal energy of the fractal static complex system. The fractal states density conservation law is identically satisfied.

References

1. Cresson, J., Adda, F.B.: Quantum derivatives and the Schrödinger equation. Chaos Solitons Fractals **19**, 1323–1334 (2004)
2. Cristescu, C.P.: Dinamici Neliniare şi Haos. Fundamente Teoretice şi Aplicaţii (in Romanian). Romanian Academy Publishing House, Bucureşti (2008)

3. Mercheş, I., Agop, M.: Differentiability and Fractality in Dynamics of Physical Systems. World Scientific Publisher, Singapore (2016)
4. Mandelbrot, B.: The Fractal Geometry of Nature. W.H. Freeman Publishers, New York (1982)
5. Notalle, L.: Scale Relativity and Fractal Space-Time: A New Approach to Unifying Relativity and Quantum Mechanics. Imperial College Press, London (2011)
6. Phillips, A.C.: Introduction to Quantum Mechanics. Wiley, New York (2003)
7. Schlichting, H.: Boundary Layer Theory. McGraw-Hill, New York (1970)

List of symbols

X
d
T
$\mathcal{C}$
$\mathcal{B}_0$
$\mathcal{B}$
$\mathcal{B}'_0$
$\mathcal{B}'$
$\widehat{\tau}_H$
$\widehat{\tau}^+_H$
$\widehat{\tau}^-_H$
$\widehat{\tau}^+_V$
$\widehat{\tau}^-_V$
$\widehat{\tau}_V$
$\widehat{\tau}_W$
$\widehat{\tau}^-_W$
$\widehat{\tau}^+_W$
$\|\cdot\|$
$|\cdot|$
$\dot{+}$
$h(M, N)$
$e(M, N)$
$\dim_{Haus}$
$\mathcal{P}_0(X)$
$\mathcal{P}_f(X)$
$\mathcal{P}(X)$
$\mathcal{P}_{bf}(X)$
$\mathcal{P}_b(X)$
$\mathcal{P}_c(X)$
$\mathcal{P}_{bc}(X)$
$\mathcal{P}_{bfc}(X)$
$\mathcal{P}_{kc}(X)$
$\mathcal{P}_k(X)$
$\mathcal{K}, \mathcal{A}$
$\mathcal{D}$
$\mathcal{I}(\mathcal{K}, \mathcal{D})$
$S(a, \varepsilon)$
$S_\varepsilon(M)$
M^-
M^+
S_{UV}
S_{LV}
D^+
D^-
$\mathcal{B}_U$
$\mathcal{B}_V$
$\mathcal{B}_{U,V_1,\ldots,V_n}$
$\mathcal{B}_{U,V}$
R'
R'_l
R'_r
R
R_r
R_l
$\mathbb{N}$
$\mathbb{N}^*$
$[0, \infty)$
$[0, \infty]$
$\mathbb{R}_+$
$\overline{\mathbb{R}}_+$
$U^H_+(M, \varepsilon)$
$U^H_-(M, \varepsilon)$
$\mathcal{F}$
U^W_-
U^W_+
$\underset{n\to\infty}{Li}(A_n)$
$\underset{n\to\infty}{Ls}(A_n)$
$\liminf\limits_{n\to\infty}$
$\limsup\limits_{n\to\infty}$
$\mu, \mu^*, \overline{\mu}, \nu, m$
$|\mu|$
$\sup$
$\inf$
$\lim$
$\max$
$\dot{\to}$
$A_n \searrow A$
$A_n \nearrow A$
$\overline{M}$
cM
NA
DP
SP
S_T
$\tau, \widetilde{\tau}$
$\widetilde{\tau}_l, \widetilde{\tau}_r$
G_δ
R^H
R^H_l
R^H_r
ρ
Ψ
$< \cdot, \cdot >$
$\mathcal{A} \times \mathcal{A}$
$\widehat{\frac{d}{dt}}$
$\partial^l, \partial_t, \partial_l$
$\widehat{V}^l$
V^l_D, V^l_F
ε_{ilk}
$\mathcal{F}(X)$
$\mathbb{C}$

A. Gavriluţ et al., *Atomicity through Fractal Measure Theory*,
https://doi.org/10.1007/978-3-030-29593-6

Index

A. Gavriluţ et al., *Atomicity through Fractal Measure Theory*,
https://doi.org/10.1007/978-3-030-29593-6

Printed in the United States
By Bookmasters